Solar Photovoltaic System Modelling and Analysis

Design and Estimation

Published 2024 by River Publishers
River Publishers
Alsbjergvej 10, 9260 Gistrup, Denmark
www.riverpublishers.com

Distributed exclusively by Routledge
605 Third Avenue, New York, NY 10017, USA
4 Park Square, Milton Park, Abingdon, Oxon OX14 4RN

Solar Photovoltaic System Modelling and Analysis / by T. Mariprasath, P. Kishore, K. Kalyankumar.

Routledge is an imprint of the Taylor & Francis Group, an informa business

ISBN 978-87-7004-090-7 (paperback)

ISBN 978-10-4002-308-2 (online)

ISBN 978-1-003-47589-7 (ebook master)

A Publication in the River Publishers series
RAPIDS SERIES IN POWER

While every effort is made to provide dependable information, the publisher, authors, and editors cannot be held responsible for any errors or omissions.

Solar Photovoltaic System Modelling and Analysis

Design and Estimation

T. Mariprasath

K.S.R.M College of Engineering (Autonomous), India

P. Kishore

IIITDM Kurnool, India

K. Kalyankumar

K.S.R.M College of Engineering (Autonomous), India

NEW YORK AND LONDON

Contents

Preface

The purpose of this book is to provide elaborate information about solar photovoltaic (PV) system modelling and analysis. The requirement for electrical energy has increased in a rapid manner due to sustainable development and the socioeconomic status of people. Hence, power industries are enhancing the power generation capacity of generating units to satisfy consumer requirements. The primary resource for conventional power plants is fossil fuels, which are extracted from the purification process of petroleum products. As per statistics, the fossil fuel depletion rate is rapidly increasing, which harms the environment. Therefore, renewable energy resources are getting more attention because they are endlessly available in the earth's atmosphere.

Globally, the electrical energy extraction rate from solar PV is rapidly increasing compared to all other renewable energy resources. As a result, technological advancement causes the cost per watt of solar PV to gradually decrease every year. The challenge of solar PV is that efficiency is low. Therefore, to improve efficiency, major technologies are incorporated, such as boost converters, maximum power point tracing techniques, solar PV cleaners, irradiation tracing systems, and adjusting the tilt angle of the panel. Around the globe, there is a huge opportunity to improve the energy yield potential of solar PV plants.

The book has seven chapters. Chapter 1 covers the importance of renewable energy resources and their types, as well as their growth rate. In Chapter 2, we go over the fundamentals of how solar PV systems produce electrical energy before examining various solar PV panel types and their significance. In Chapter 3, different types of solar PV configurations and their structures are described. In Chapter 4, the role of charge controllers, different types of charge controllers, and a brief discussion on conventional and bioinspired MPPT techniques. In Chapter 5, a brief discussion is given on the buck and boost converters, the modelling of boost converters, and how boost converters behave under steady and variable irradiations. In addition, with a modified boost converter, performance is analyzed. Chapter 6 describes the necessity of hybrid power plants and grid-connected solar PV systems. In Chapter 7, we provide information about the size of solar PV, which includes an estimation of the load corresponding to the size of solar PV, a boost converter, and an inverter rating for domestic applications. Hence, the book is very useful for bachelor's and master's degree students as well as research scholars.

About the Authors

Dr. T. Mariprasath received his Ph.D. degree from the Rural Energy Centre, The Gandhigram Rural Institute (Deemed to be University), in January 2017, fully funded by the Ministry of Human Resource Development—Government of India. He has been working as an Associate Professor with the Department of EEE, K.S.R.M. College of Engineering (Autonomous), Kadapa, since June 2018. He has published in 10 science citation indexed journals and 15 Scopus indexed articles. Moreover, he has published a book with Cambridge Scholars Publishing, UK. He also has an Indian patent granted. His research interests include renewable energy engineering, renewable energy resources, micro grid, green dielectrics, and artificial intelligence.

Dr. P. Kishore received his Ph.D. in Physics with Photonics Specialization from National Institute of Technology in the year 2015. He completed his M.Sc. Physics (Electronics) and B.Sc. (M.P.Cs.) from the Sri Venkateswara University Tirupati. He has eight years of teaching experience in physics and photonics subjects. He has published 70 research articles in reputed journals and conference proceedings. He has also published four book chapters and filed one Indian IPR patent. Currently he is working as an Assistant Professor of Physics in the department of Sciences at Indian Institute of Information Technology Desing and Manufacturing Kurnool, Andhra Pradesh India. He works in the areas of optics, photonics, renewable energies and allied areas.

Mr. K. Kalyankumar is pursuing a Ph.D at Yogi Vemana University, Andhrap Pradesh, India, He has been working as an Assistant Professor with the Department of EEE, K.S.R.M. College of Engineering (Autonomous), Kadapa, since December 2006. He has published five Scopus indexed articles and four book chapters. His research interests include power systems, electrical machines and power electronics.

1

Renewable Energy

1.1 Renewable Energy

Renewable energy means energy derived from natural sources that are replenished at a higher rate than they are consumed. Sunlight and wind are the best examples of such sources that are constantly being replenished. Renewable energy sources plentifully available all around us. On the other hand, fossil fuels, coal, oil and gas, are non-renewable resources that take hundreds of millions of years to form. In addition, fossil fuels when burned to produce energy, produce harmful greenhouse gas emissions, such as carbon dioxide, etc. However, generating renewable energy creates far lower emissions than burning fossil fuels. Transitioning from fossil fuels, which currently account for the lion's share of emissions, to renewable energy is key to addressing the climate crisis. Day to day renewables have become cheaper in most countries, and generate three times as many jobs than fossil fuels.

The use of fossil fuels consumes non-renewable resources, which means that these resources will eventually run out, at which point their extraction will either be prohibitively expensive or environmentally detrimental. On the other hand, the many different types of renewable energy resources, such as wind and solar energy, are continuously being replenished and will never run out. When coal is burned for energy, a wide variety of different kinds of particulate matter are produced, all of which contribute to air pollution. When one considers the various non-financial costs of using fossil fuels, such as pollution, climate change, and the effect on biodiversity, renewable energy is actually much more efficient than fossil fuels. This is because pollution causes climate change, which in turn impacts biodiversity. One of the drawbacks of using renewable sources of energy is that it is difficult to generate significant quantities of electricity comparable to that which is produced by conventional generators that run on fossil fuels [1], [2]. The supply of power for renewable energy sources is frequently dependent on the elements. Rain is necessary to fill dams and furnish hydroelectric generators with flowing water. The fact that most forms of alternative energy are, on average, more expensive than energy derived from fossil fuels is perhaps the most significant disadvantage of using renewable sources of power. Because of this, there has not been as much of a shift toward the use of sustainable sources of energy as there has been toward the use of clean energy [3]–[5].

The majority of the energy that is renewable originates from the sun, either directly or indirectly. Sunlight and solar energy can be used directly for a variety of purposes, including the heating and lighting of homes and other structures; the generation of electricity; solar cooking; solar water heating; and a wide range of other commercial and industrial applications. Wind turbines are able to harness the power of the wind because the heat from the sun is also what powers the wind. Then, the heat from the sun and the breeze cause water to evaporate; later, this water vapor condenses into precipitation in the form of rain or snow, which then flows into rivers or streams. Renewable energy sources have quickly become a preferred alternative to conventional methods of producing electricity in settings where this type of production simply isn't feasible. In this day and age, we need to focus on renewable resources in order to satisfy the desire for electricity. Wind power is widely recognized as one of the most promising areas in the field of sustainable energy. Wind electricity is a resource that is both clean and environmentally friendly. Because of the variations in wind speed brought on by the changing seasons, the production of electricity from wind turbines can be quite unpredictable. The wind generator that is coupled with the wind turbine will generate voltage that is variable as well as a frequency that is variable. As a result, the objective of the endeavor is to keep the voltage and frequency on the output side at a constant level [6], [7].

The majority of agricultural irrigation is a complex interplay that involves the consumption of renewable energy, the use of water, the conditions of the market, as well as the application of experience and knowledge to guarantee the best design for irrigation applications. The market for agricultural products is shifting at a breakneck pace, making it impossible for farmers to depend on the technology and practices of today. Induction motors with three phases are typically found in agricultural applications. It is possible for it to be either a three-phase or a one-phase induction motor, depending on the application. One-phase induction motors are suitable for use in residential settings, while three-phase and one-phase motors are suitable for use in commercial settings.

They are maintenance motors in contrast to dc motors and synchronous motors due to the absence of brushes, commutations, and slip rings. The construction of these motors is simple and robust, and they have a lower cost than other types of motors because they do not have these components [8].

The use of fossil fuels consumes non-renewable resources, which means that these resources will eventually run out, at which point their extraction will either be prohibitively expensive or environmentally detrimental. On the other hand, the many different types of renewable energy resources, such as wind and solar energy, are continuously being replenished and will never run out. When coal is burned for energy, a wide variety of different kinds of particulate matter are produced, all of which contribute to air pollution. When one considers the various non-financial costs of using fossil fuels, such as pollution, climate change, and the effect on biodiversity, renewable energy is actually much more efficient than fossil fuels. This is because pollution causes climate change, and pollution causes climate change, which in turn impacts biodiversity. One of the drawbacks of using renewable sources of energy is that it is difficult to generate significant quantities of electricity comparable to that which is produced by conventional generators that run on fossil fuels. The supply of power for renewable energy sources is frequently dependent on the elements. Rain is necessary to fill dams and furnish hydroelectric generators with flowing water. The fact that most forms of alternative energy are, on average, more expensive than energy derived from fossil fuels is perhaps the most significant disadvantage of using renewable sources of power. Because of this, there has not been as much of a shift toward the use of sustainable sources of energy as there has been toward the use of clean energy.

1.2 Non-Conventional Power Plants

1.2.1 Wind power plant

In order to generate electricity, a wind farm must first collect the dynamic energy of the wind and then transform it into electrical energy. The wind turbine, the rotor, the gearbox, the generator, and the management mechanism make up the fundamental elements of a wind power facility. A wind turbine comprises of a tall structure that supports a rotor that has two or three blades. The number of blades can vary. The wind is captured by the blades, and their construction allows them to transform the wind's kinetic energy into rotational energy. The rotor spins as a result of the wind's force and powers a gearing, which in turn causes an increase in the rotor's rotational speed. The rotor's kinetic energy is transferred to an attached generator, which then produces usable electrical current. The generator links to the transmission. The generator generates electrical power, which is then sent to a transformer. The transformer raises the voltage of the electricity, and then the electricity goes to the power system. The management system of the wind power plant acts as a measure for the performance of the turbine, ensuring that it continues to function in an effective and secure manner at all times. In addition to this, it alters the inclination of the blades so that the amount of power generated is maximized in accordance with the speed and direction of the wind [9], [10]. However, the growth of thermal power plants, as shown in Figure 1.1, is comparable to the growth of wind power plants, as shown in Figure 1.2.

Figure 1.1: Thermal power plant growth.

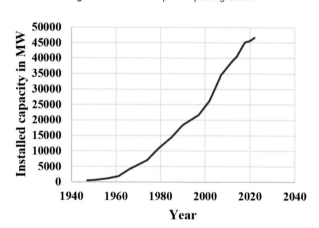

In recent years, there has been a substantial increase in the number of wind power plants, which can be attributed to a number of factors, including technological breakthroughs, cost reductions, and government policies that encourage the use of renewable energy sources. The expansion of wind power facilities has been characterized by the following main trends. Capacity that has been installed: The Global Wind Energy Council estimates that by the end of 2021, the total installed capacity of wind power facilities across the globe will have reached 743 GW. This represents a significant increase from the meagre total of 18 GW in the year 1999. At the moment, the market for wind power in China is the biggest in the world. China is followed by the markets in the United States, Germany, India, and Spain. The price of electricity generated by wind has been gradually declining over the past few years, making it an increasingly competitive alternative to more conventional forms of electricity production. In some areas, the cost of electricity generated by wind power is already lower than

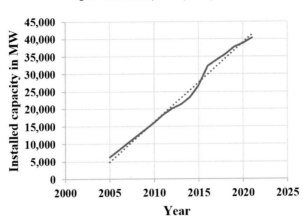

Figure 1.2: Wind power plant growth.

that generated by natural fuels. It is anticipated that this pattern will continue as technological advancements continue and efficiencies of scale are attained. This is particularly the case in Europe, where they have the potential to provide a substantial source of sustainable energy, and offshore wind power plants are gaining a lot of popularity and are expected to continue doing so in the near future. In 2010, offshore wind power facilities had a total operational capacity of just 2 GW, but by the end of the year 2020, that number had increased to 35 GW [11], [12].

1.2.2 Biomass power plant

A biomass power plant is a form of renewable energy plant that generates electricity by burning organic materials such as timber, crop residue, and waste products from forests, fields, and industries. Timber and agricultural crops are two examples of organic materials that can be burned in a biomass power plant. The basic idea behind a biomass power plant is comparable to that of a traditional thermal power plant; however, there are several important distinctions between the two. The following is a list of the fundamental elements that make up a bioenergy power plant. The organic materials that are used as fuel are stacked in large heaps or housed in silos close to the power station. In order to promote more efficient combustion, the fuel is generally chipped or ground into minute fragments. The fuel is moved from the storage area to the furnace using a conveyor belt or another mechanical device. This is part of the fuel management process. The fuel is consumed in a combustion chamber, and the resulting heat is used to heat the water in a furnace, which ultimately results in the production of steam. The steam turns a turbine, which ultimately results in the production of electricity. The combustion process results in the discharge of various gases and particles into the environment, including particulate matter, nitrogen oxides, and sulfur dioxide. Scrubbers, electrostatic precipitators, and selective catalytic reduction systems are just some of the technologies that are utilized in the process of reducing and controlling these pollutants [13], [14].

Power production and distribution: After the turbine has finished generating electricity, the current is sent to a transformer, which boosts the voltage of the current before sending it on to the power infrastructure. Since the biological materials that are used as fuel in biomass power plants can be replenished over time, this type of power plant is regarded as a source of sustainable energy. However, the sustainability of biomass as a fuel source is contingent on the type of biomass that is utilized as well as the harvesting practices and transportation methods that are utilized. In addition, biomass power plants can have a significant effect on air quality in the surrounding area; therefore, it is essential that pollutants be tightly monitored in order to minimize the damage they cause to the environment [15], [16].

1.2.3 Ocean power plant

A form of sustainable energy plant that generates electricity by converting the kinetic energy of ocean waves or tides is referred to as an ocean power plant. This type of plant is also referred to as a wave energy conversion system or a tidal power plant. Ocean power plants can take a few different forms, but the basic idea behind all of them is the same: to transform the mechanical energy of ocean waves or swells into electrical energy. The following is a list of some of the key elements of a water power plant: A wave energy converter is a piece of equipment that transforms the kinetic energy of waves into a usable form of energy, typically electricity. Wave energy can be converted in a number of different ways, depending on the specific design of the device, such as with point absorbers, oscillating water columns, or overtopping devices. A turbine that is propelled by the movement of tidal currents is referred to as a tidal turbine. The tidal turbine is generally installed on the bottom of the ocean, and it spins as the tidal currents travel past it. This rotation drives a generator, which in turn produces electricity. The electrical energy that is generated by the wave energy converter or the tidal turbine is then transferred to an electrical system, where it is conditioned and transformed so that it can fulfil the requirements of the power infrastructure. The power plant located in the water sends the electricity it generates to a transformer, which is then connected to the power system. From there, the electricity is disseminated to various consumers. The establishment of ocean power plants is still in its infancy, and there are presently only a handful of commercial-scale ocean power plants operational. However, because ocean waves and swells are a reliable and continuous source of energy, there is a substantial opportunity for ocean power to serve as a source of sustainable energy in the future. Ocean power also has the benefit of being situated near heavily inhabited shoreline regions, which have a high demand for electricity and can take advantage of the proximity of ocean power plants. Despite this, there are still substantial hurdles to get over in terms of the expense of the technology, its effect on the environment, and other factors [17], [18].

1.2.4 Hydro power plant

This is a form of sustainable energy plant known as a hydroelectric plant, and its primary function is to produce electricity by capturing the kinetic energy of flowing water. The size of hydropower facilities is not a determining factor in their location; they can be constructed on rivers, streams, or even man-made waterways. The following is a list of the fundamental parts that make up a hydroelectric power plant: A reservoir can be created by constructing a dam or weir, which are both types of structures that increase the water level in a waterway or stream. The dam or the weir is used to store the water until the time comes when it is required to generate electricity. An entrance or other aperture within the dam or the weir that enables water to move into the power facility. A conduit that transports water from the inflow to the generator in the form of a pipeline. A machine that rotates as a result of the power exerted by water as it passes through it. The generator, which is connected to the turbine, is the component that generates the electricity. A component that takes the mechanical energy that is produced by the turbine and transforms it into electrical energy. A piece of equipment that raises the frequency of the electricity generated by the generator so that it can be fed into the public power supply. Cables or lines that connect the hydroelectric plant to the power infrastructure and transport electricity from the plant to the grid. The water that is used to generate electricity in hydropower facilities is continually refilled by rain and precipitation, making this type of power generation a source of sustainable energy. In addition, hydropower is a versatile source of electricity because the quantity of power generated by hydroelectric dams can be adjusted to meet fluctuating consumer needs. However, the construction of large hydroelectric plants can have a substantial influence on nature, including the relocation of wildlife and the modification of waterway environments. These effects can be avoided by minimizing the size of the plants [19]–[21].

Figure 1.3: Hydro power plant potential.

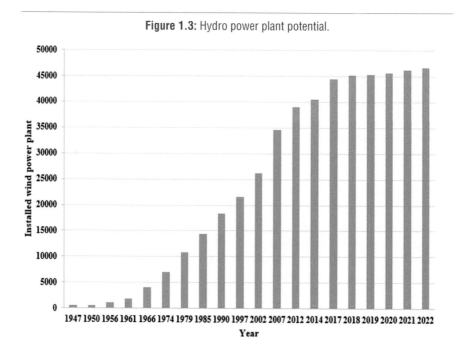

Hydropower plant construction has been consistent over the years, with hydropower being one of the oldest and most established types of renewable energy. Here are some patterns in hydropower plant development: The International Hydropower Association estimates that the global capacity of hydropower plants will be 1408 GW in 2020, with an estimated 5306 Tw of electricity produced per year. This indicates a 60% increase in capacity since 2000.

China, Brazil, Canada, the United States, and Russia are the top five nations in terms of installed hydropower capacity. China alone accounts for more than one-third of total installed hydropower potential worldwide. Large-scale hydropower projects are presently under construction or in the planning stages, including the Grand Ethiopian Renaissance Dam, the Inga III Dam in the Democratic Republic of the Congo, and the Baku Dam in Malaysia. Small-scale hydropower projects: In addition to large-scale hydropower projects, there is increasing interest in small-scale hydropower projects built on small rivers or streams. These projects are frequently used to provide power to remote communities or small companies. Pumped-storage hydropower plants are becoming more essential as a way of balancing the grid and storing excess electricity generated by intermittent renewable energy sources such as wind and solar. When demand is high, these plants use surplus electricity to pump water from a lower reservoir to a higher reservoir, and then release the water through turbines to produce electricity. The construction of hydropower plants is anticipated to increase in the coming years, as many nations seek to increase their renewable energy capacity and reduce their reliance on fossil fuels. The environmental effect of large-scale hydropower projects, on the other hand, is still a worry, and there is a need to balance the benefits of hydropower with the potential impact on ecosystems and local communities [22].

1.2.5 Geothermal energy

Geothermal energy is a form of sustainable energy that generates power by harnessing heat from the earth's core. The earth's core is incredibly heated, and geothermal power facilities can reach this heat in certain regions. The following is how geothermal energy works: The geothermal reserve is a place where the earth's heat is close to the surface, typically near the meeting points of the earth's geological zones. Water is heated by the earth's heat and turns into vapor in these regions. To transport the vapor and heated water to the surface, a producing well is dug into the geothermal pool. Steam or heated water is used to fuel an engine, which is linked to a generator, which generates energy. The chilled water is then rejected back into the earth to keep the geothermal pool pressure constant. Geothermal power plants are classified into three types: dry steam power plants, flash steam power plants, and binary cycle power plants. The sort of power facility used is determined by the geothermal reservoir's pressure and temperature [23], [24].

Geothermal energy is a dependable and constant form of sustainable energy because the heat from the earth's core is virtually limitless. Geothermal energy can also be used to heat and chill structures using ground-source heat exchangers. The growth of geothermal power facilities, on the other hand, can have environmental consequences, such as the emission of carbon gases and the possibility for earthquake action in regions where geothermal energy is harvested. Geothermal energy is a tiny but increasing sustainable energy source. Here are some patterns in geothermal energy growth: The International Renewable Energy Agency (IRENA) estimates that the global capacity of geothermal power facilities will be 14.9 GW by the end of 2020. This reflects a 25% rise since 2010. The United States, Indonesia, the Philippines, Turkey, and New Zealand have the most operational geo thermal power.

These nations have advantageous geothermal resources and have invested in geothermal energy for decades. Geothermal resources are being explored and developed in a number of nations, including Mexico, Kenya, Iceland, and Japan. These nations have discovered substantial geothermal deposits and are engaging in the construction of new geothermal power facilities. Geothermal energy can also be used for direct uses like powering and cooling structures or manufacturing operations. Geothermal energy direct-use uses are more prevalent than geothermal electricity production and can be found in many nations around the globe. Important technical advances in geothermal energy have occurred in recent years, including the creation of novel digging methods and the use of binary cycle power plants, which can run at lower temps than conventional geothermal power plants [25], [26], and [27].

Regardless of its promise, geothermal energy confronts several barriers to development, including high setup expenses, a lack of suitable geothermal resources, and the possibility of negative environmental consequences. However, increasing consumer appetite for green energy and the continuing creation of new technologies are anticipated to propel the geothermal power industry forward.

CHAPTER

2

Photovoltaic Systems

2.1 Photovoltaic System

A power system that is intended to supply usable power by means of photovoltaics is referred to as a photovoltaic system, also known as a PV system or a solar power system. It is an arrangement of several components: solar panels, which take in sunlight and transform it into electricity; a solar inverter, which converts the electric current from direct current to alternating current; mounting, cabling, and other electrical accessories, which are used to set up a functioning system; and so on. In addition to this, it might incorporate a solar tracking system in order to boost the system's overall performance and have an incorporated battery solution because the cost of energy storage devices is anticipated to fall in the near future. Strictly speaking, a solar array consists of nothing more than a collection of solar panels, which is the visible portion of the photovoltaic (PV) system. It does not include any of the other hardware, which is commonly referred to as the balance of system (BOS). Moreover, PV systems convert light directly into electricity and should not be confused with other technologies, such as concentrated solar power or solar thermal, which are used for heating and cooling systems. These technologies range in size from small, rooftop-mounted or building-integrated systems with capacities from a few to several tens of kilowatts, to large utility-scale power stations of hundreds of megawatts. PV systems convert light directly into electricity. These days, the vast majority of PV systems are connected to the grid, while only a small percentage are off-grid or stand-alone systems [28], [29].

Through this attempt, we will be able to lessen our impact on the environment. The electricity that is fed to the motors is generated through the use of renewable energy resources, specifically the existence of solar energy. The primary objective is to cut down on environmental pollution while simultaneously increasing the use of solar power (photovoltaic cells) in place of electricity generated by fossil fuels. After a natural catastrophe has occurred, the power grid may be rendered inoperable for a period of time. Therefore, the utilization of these forms of energy, such as wind energy, hydraulic energy, and solar energy, is essential for the production of electrical power. Solar energy has emerged as a potentially useful alternative energy source due to the many benefits it offers, including its abundance, absence of pollution, and capacity to be continuously renewed. Using solar energy is beneficial for a number of reasons, including the fact that the lifespan of a solar panel is significantly longer than the lifetime of any other type of energy source. People who live in more remote regions should absolutely invest in solar charging systems because they are extremely useful and essential to their daily lives. Because solar cells are utilized in the generation of electricity rather than hydraulic generators, the sun is a relatively inexpensive supply of this resource. We make use of the sun as a natural source of electricity in modern times. Solar energy is a limitless resource. Governments are attempting to implement the use of solar panels as an energy source in rural and semi-urban areas for the purpose of powering street lights; however, the battery that is used to store the power is being negatively impacted as a result of overcharging and discharges [31], [32].

Solar power not only has a promising future in terms of how it can be utilized, but it also stands to play a significant part in easing the current energy crisis and cutting down on emissions in the environment. A charge controller is a component that is necessary for the majority of different types of power systems, including those that charge batteries, wind, hydro, fuel, or the utility infrastructure [33]. It is designed to ensure that the battery is adequately nourished and protected over an extended period of time. A regulator that is installed in between the solar arrays and the batteries is referred to as a charge controller. Regulators for solar systems are built to keep the batteries charged to their maximum capacity while preventing them from being overcharged. A solar charge controller has many potential applications in a variety of fields. For instance, it could be utilized in a solar house system, hybrid systems, a solar water pump system, and other similar applications. During this process, solar panels use an electrochemical process that is also known as the photovoltaic process to transform the energy from the sun into usable electrical energy. Through the use of a diode and a fuse, energy is transferred from the solar panel to the battery so that it can be retained. When the battery is discharged, the chemical energy is transformed into electrical energy, which can then be used to illuminate electric appliances or assist in the process of pumping water from the ground. The energy that is stored in the battery can be used even when there is no sunlight. Due to the fact that it is the primary component in a solar power charge controller, it is necessary to prevent the battery from being overcharged, from going into a deep discharging mode while dc loads are being used, and from going into an under-voltage state [34], [35].

There are two primary classifications that can be applied to photovoltaic solar systems: grid-connected solar systems and off-grid solar systems, which are also referred to as standalone or solitary solar systems. Through the use of an inverter, the grid-connected devices deliver the electricity that was generated by solar panels to the grid. When there is a demand for electricity during the night or other times of the day when there is little sunlight, the energy is drawn back from the system. In isolated systems, surplus electricity is typically stored in batteries during the day, and these batteries are then used to power appliances during times when the energy produced by photovoltaic panels is insufficient. Solar regulators, which are also referred to as charge controllers, play an essential part in the operation of standalone solar power systems. Their objective is to ensure that the batteries are operating at their maximum potential, primarily to prevent the batteries from being overcharged (by turning off the solar panels when the batteries reach their capacity) and from being discharged to an excessively low level. The lifespan of a battery is dramatically shortened when it is overcharged and then deeply discharged. Because the battery is such an expensive component of a solar home system, it is essential to take precautions to prevent it from being overcharged or completely drained of its charge [36], [37].

In this sense, a charge controller is an extremely important component for ensuring the battery's safety. A solar charge controller, a battery, and a photovoltaic panel make up the components of this device. Batteries are used to retain the energy harvested from the sun. A solar charge controller is a device that manages the voltage and current that flow from photovoltaic panels (solar panels) to an electrical storage device (battery). The charge controller is a switching mechanism that regulates how much the battery is charged as well as how much it is discharged. Because of this, the batteries will be protected from injury, which will result in a longer lifespan for the batteries.

A photovoltaic system includes a charge controller, batteries, and a power converter. It also includes a PV or solar panel, which is called a module, as shown in Figure 2.1. The photovoltaic panel, also known as a solar module or array, is responsible for converting the energy from the sun into direct current electrical energy. A battery can be charged thanks to the charge controller, which regulates the DC electrical voltage and current that are generated by a

Figure 2.1: Block diagram of photovoltaic (PV) system.

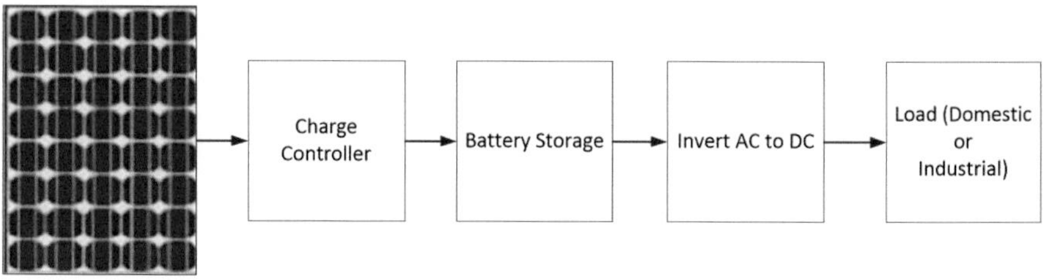

| Charge Controller | → | Battery Storage | → | Invert AC to DC | → | Load (Domestic or Industrial) |

photovoltaic panel (module) or solar array. The battery is responsible for storing the direct current (DC) electrical energy so that it can be utilized even when there is no accessible solar energy (night time, cloudy days, etc.). Direct current (DC) loads are able to receive power straight from the photovoltaic module, solar panel, or battery. The DC power that is generated by the PV or solar panel (module) and stored in the battery is changed into alternating current (AC) power by the converter so that AC loads can be powered.

2.2 Photovoltaic Power Generation

Figure 2.2: Photovoltaic effect.

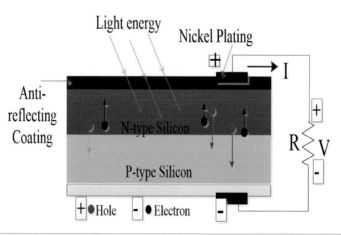

A phenomenon referred to as the photovoltaic effect is responsible for solar cells' ability to convert sunshine into usable electricity. Solar cells are also known as photovoltaic cells. The production of energy by specific materials, such as silicon, when they are subjected to the action of sunshine is referred to as the photovoltaic effect. Silicon is a type of substance known as a semiconductor, and a solar cell is made up of layers of silicon. The upper layer of the solar cell is made up of a very thin layer of phosphorus-doped silicon, as shown in Figure 2.2, which has an overabundance of electrons. This layer is what makes up the solar cell. The base layer is composed of a more substantial layer of boron-doped silicon, which has a lower number of electrons than the other layers. When light from the sun strikes a solar cell, photons from the light are taken in by the silicon crystals that are located in the solar cell's upper layer. Because of this, electrons in the silicon are dislodged from their atomic bonds, which results in a movement of electrons between the upper and lower levels of the solar cell. The movement of electrons results in the production of a direct current (DC), which is then routed to a converter, where it is changed into an alternating current (AC), which is the type of electricity that can be used to power buildings like houses and businesses. The holes and electron production on respective materials as shown in Figure 2.2 [38], [39].

Solar panels can be made in a wide range of shapes and power outputs by connecting solar cells in either series or parallel connections. A solar panel's capacity to produce usable electricity is directly proportional to the number of solar cells it contains. In general, solar cells offer a clean and sustainable source of electricity that, when utilized, can contribute to the reduction of pollutants that contribute to global warming and the promotion of energy independence.

2.3 Monocrystalline Silicon Solar Cell

As you can see Figure 2.3, solar cells that are made of monocrystalline silicon (mono-Si), which is also known as single-crystalline silicon (single-crystal-Si), are quite easy to identify because they have an even coloring and uniform appearance on the outside, which indicates that they are made of high-purity silicon.

Silicon crystals, which have a cylindrical form, are the raw material that go into making monocrystalline solar cells. Silicon wafers are created by cutting four sides off cylindrical ingots to produce monocrystalline solar panels, as shown in Figure 2.2. This is done to improve the efficiency of a single monocrystalline solar cell as well as to reduce the associated costs. One of the most helpful ways to differentiate between monocrystalline and polycrystalline

Figure 2.3: Monocrystalline solar module.

solar panels is by looking for precisely rectangular cells that do not have any rounded edges. Solar panels that are built from monocrystalline silicon have the highest efficiency rates because the silicon used to make them is of the highest quality. In general, monocrystalline solar panels have efficiency values that range between 15% and 20%. SunPower's solar panels currently have the greatest efficiency rating available on the market in the United States. Panel conversion efficiencies of up to 20.1% are provided by their E20 family of products. Update (April 2013): SunPower has now launched the X-series, which boasts an efficiency rate of 21.5%, shattering all previous records. Monocrystalline silicon solar panels are space efficient. When compared to other kinds of solar panels, these ones have the capability of producing the highest amounts of power while also requiring the least amount of space. Solar panels made of monocrystalline silicon generate up to four times the quantity of electricity as solar panels made of thin-film silicon. Solar panels made of monocrystalline silicon have the greatest lifespan. The majority of solar panel manufacturers include a warranty of 25 years on their monocrystalline solar panels, and have a tendency to perform better in low-light situations compared to similarly rated polycrystalline solar panels [41], [42].

Solar panels made of monocrystalline silicon are the most expensive. For some householders, the purchase of a solar panel that is constructed out of polycrystalline silicon (or thin-film technology, in certain instances) may prove to be the most beneficial option financially. It is possible for the complete circuit to be disrupted if even a portion of the solar panel is obscured by shade, dirt, or snow. If you anticipate that coverage will be an issue, you might want to consider switching to micro-inverters from central string inverters. Because of micro-inverters, the solar array as a whole will not be negatively impacted if there are shadowing problems with just one of the solar panels. The production of monocrystalline silicon is accomplished through the Kochanski procedure. It produces massive ingots that are cylindrical in shape. In order to produce silicon wafers, the ingots must first have all four edges cut away. A sizeable portion of the silicon that was initially extracted is lost as trash. In warmer climates, the efficiency of solar panels made of monocrystalline silicon tends to be higher. The performance of monocrystalline solar panels is less affected by an increase in temperature than the performance of polycrystalline solar panels. Temperature is not a problem for the vast majority of homeowners [43].

2.4 Polycrystalline Silicon

Polycrystalline silicon, also known as polysilicon (p-Si) and multi-crystalline silicon (mc-Si), was first used to make solar panels in 1981, when they were first introduced to the market. Polycrystalline solar panels, as shown in Figure 2.4, as opposed to monocrystalline solar panels, do not require the Kochanski procedure in order to be manufactured. The raw silicon is melted down and poured into a square meld before being allowed to cool and then being cut into wafers that are precisely square [44], [45].

The manufacturing procedure for polycrystalline silicon is less complicated and less expensive than other methods. In comparison to monocrystalline, polycrystalline produces a smaller quantity of waste silicon as shown in Figure 2.4. Solar panels made of polycrystalline silicon typically have a slightly lower heat threshold than solar panels made of monocrystalline silicon. In a scientific sense, this indicates that their performance is marginally lower than that of monocrystalline solar panels when exposed to high temperatures. Solar panels can have their efficiency reduced and their lifespans shortened when exposed to heat. The majority of homeowners, however, do not need to worry about this impact because it is relatively insignificant [46].

The effectiveness of polycrystalline solar panels ranges between 13% and 16% on average. Solar panels made of polycrystalline silicon are not quite as efficient as solar panels made of monocrystalline silicon due to the lower purity of the silicon. In addition, requirement of space is high for producing same output as that of mono crystalline panels. If you want to generate the same amount of electrical power as you would with a solar panel made of monocrystalline silicon, you will typically need to cover a much bigger surface area. However, this does not necessarily imply that the performance of monocrystalline solar panels is superior to that of polycrystalline silicon-based panels. When compared to solar panels made of polycrystalline silicon, which have a speckled blue color, monocrystalline and thin-film solar panels, which have a more uniform look, are generally considered to be more aesthetically appealing [47].

2.5 Solar Cells Made of a Thin Film (TFSC)

The production of thin-film solar cells is essentially accomplished by depositing one or more thin layers of photovoltaic material onto a substrate. This can be done multiple times. Thin-film photovoltaic cells is another name for these types of cells, shown in Figure 2.5. One way to classify the many varieties of thin-film solar cells is according to the photovoltaic substance that is deposited onto the substrate in the following order: amorphous silicon, also known as a-Si; cadmium telluride, also known as CdTe; copper indium gallium selenide, also known as CIS/CIGS; organic photovoltaic cells (OPC). Thin-film module prototypes, as shown in Figure 2.5, have achieved efficiencies ranging from 7–13%, depending on the technology, while manufacturing modules have an efficiency of approximately 9%. It is anticipated that future

Figure 2.4: Polycrystalline solar module.

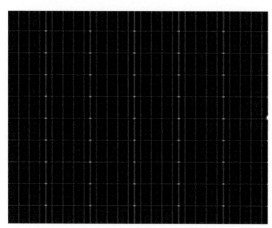

modular efficiencies will increase to somewhere between 10% and 16%. From 2002 to 2007, the market for thin-film PV expanded at a pace of 60% per year on average. In 2011, nearly 5% of the photovoltaic module shipments made in the United States to the residential sector were based on thin-film technology [48], [49].

Figure 2.5: Thin film solar panels.

Production on a large scale is not complicated. Because of this, their production could be less expensive than that of solar cells built on crystalline material. Because of how similar they are to one another, they come across as more appealing. It is possible to make them adaptable, which paves the way for a wide range of new possible applications. High temperatures and being in the shade have less of an effect on the effectiveness of solar panels. It may make sense to use thin-film solar panels in settings where there is ample room for their installation.

In most residential settings, thin-film solar panels do not offer a great deal of utility and are not recommended for use. They are inexpensive, but they take up a significant amount of space. When compared to thin-film solar panels, SunPower's monocrystalline solar panels generate up to four times the amount of electricity in the same amount of space as thin-film solar panels. Low space-efficiency also means that there will be an increase in the price of PV equipment, such as support structures and cables. Because thin-film solar panels tend to degrade more quickly than monocrystalline and polycrystalline solar panels, the warranties that accompany them are generally for shorter durations of time.

2.6 Solar Cells Made out of Amorphous Silicon (a-Si)

Solar cells based on amorphous silicon have traditionally only been utilized for low-scale applications such as those found in pocket calculators due to the minimal amount of electrical power that can be generated by these cells, as shown in Figure 2.6. Recent developments, on the other hand, have also made them more appealing for use in a variety of large-scale applications. Multiple layers of amorphous silicon solar cells can be stacked on top of one another during the manufacturing process, known as "stacking." This results in higher efficiency rates, which are generally in the range of 6–8%. In order to make an amorphous silicon solar cell, you only need 1% of the silicon that is used to make a crystalline silicon solar cell. Stacking, on the other hand, incurs additional costs [50], [51].

Figure 2.6: Cadmium telluride (CdTe) solar cells.

Cadmium telluride, as shown in Figures 2.6 and 2.7 is the only thin-film solar panel technology that has surpassed the cost-efficiency of crystalline silicon solar panels in a substantial portion of the market. This is because crystalline

Figure 2.7: Cadmium telluride solar module.

silicon solar panels are made of a much thicker layer of material. In general, solar panels that are built on cadmium telluride have an efficiency that falls somewhere in the range of 9–11%. Around the globe, First Solar has successfully installed more than 5 GW of thin-film cadmium telluride solar panels. The world record for the efficiency of a CdTe PV module currently sits with the same business at 14.4%.

2.7 Copper Indium Gallium Selenide (CIS/CIGS)

In terms of effectiveness, CIGS solar cells have demonstrated the most promising potential when compared to the other thin-film technologies described above. These solar cells have lower concentrations of the hazardous element cadmium compared to the CdTe solar cells that are also available. In 2011, Germany became the first country in the world to begin commercial manufacture of flexible CIGS solar panels. Solar panels made of CIGS material generally have efficiency rates that fall somewhere between 10% and 12%. There are a lot of different types of thin-film solar cells that are still in the research and testing phases. A few of them have a tremendous amount of untapped potential, and it is highly possible that we will see more of them in the years to come.

CHAPTER

3

Types of Solar Photovoltaic Systems

3.1 Types of Solar Photovoltaic Systems

There are three primary categories of solar photovoltaic (PV) systems with accompanying battery storage: grid-tied, grid/hybrid, and off-grid. They each have their own set of benefits and drawbacks and, ultimately, the decision will boil down to the customer's existing energy source and the outcomes they hope to achieve from using the system.

3.1.1 Grid-tied systems

Figure 3.1: PV direct interconnect with utility grid.

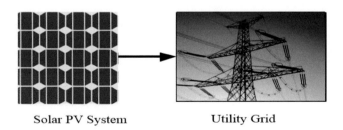

Solar PV System   Utility Grid

A solar installation that is considered to be "grid-tied" is one that employs a conventional grid-tied inverter but does not contain any batteries for energy storage, as shown in Figure 3.1. Customers who are already connected to the power and want to install solar panels on their home will find this option to be ideal. These systems may be eligible for financial assistance from the state as well as the federal government to help defray the cost of the system. Because they contain a comparatively small number of parts, grid-connected systems are not only straightforward to create but also very economical. The primary objective of a system that is connected to the grid is to reduce your monthly energy cost and reap the benefits of solar incentives. The fact that your system is rendered useless in the event that the power supply fails is one of the many drawbacks associated with this type of system. Linemen who are working on the electricity lines need to be aware that there is no source currently feeding the grid for reasons pertaining to their safety. Inverters that are connected to the grid are required to disconnect themselves automatically if they do not detect the grid. Because of this, you won't be able to provide electricity in the event of a blackout or emergency and you won't be able to store energy for later use. You also lack the ability to regulate the times at which you draw electricity from your system, such as during periods of high demand [52], [53].

3.1.2 Grid-tied system with battery back-up

Figure 3.2: Grid tied system.

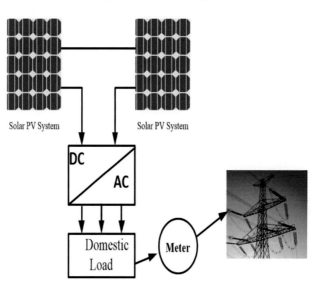

The next type of system is referred to as a grid/hybrid system and it is a system that is connected to the grid but also has a battery backup, as shown in Figure 3.2. Customers who are already connected to the grid and are aware of their desire to have battery backup are the perfect candidates for this kind of device. Customers who live in areas that frequently experience power disruptions or who simply want to be in a position to be self-sufficient in the event of a blackout are ideal candidates for this kind of system. You get the best of both worlds with this type of system because it keeps you connected to the grid, allowing you to be eligible for state and government incentives, while also reducing the amount of money you spend on your monthly utility bill. On the other hand, in the event that there is a power failure, you have a back-up. During a power outage, battery-based grid-tied systems continue to supply power and allow you to store energy for use in the event of a catastrophe. In the event that the power goes out, you have the ability to back up important loads such as lighting and appliances. Additionally, because you are able to store energy in your battery bank for later use, you are able to use energy even during periods of high demand [54].

This form of system has the drawbacks of being more expensive than standard grid-connected systems and having lower energy efficiency. Additionally, there are additional components. The incorporation of the batteries necessitates the use of a charge controller in order to ensure their safety. In addition to this, there should be a subpanel that houses the essential files that you need to make sure are backed up, as shown in Figure 3.3. The system does not provide backup for all of the loads that the home has connected to the public power infrastructure. A back-up sub-panel is used to compartmentalize critical tasks that must be operational in the event that the grid loses power. This slide will walk you through the components of a simple grid-tie system as well as the assembly process. There are just a few key components that have been added to the electricity connection that is already in place. The inverter is wired so that it connects straight to the primary service panel.

3.1.3 Off-grid system

Customers who are unable to readily connect to the grid benefit greatly from the use of off-grid systems, shown in Figure 3.4. It's possible that this is due to the location of the power plant, or the high expense of transporting the power there. A person who is currently connected to the grid should consider carefully whether or not it is in their best interest to completely disconnect and install an off-grid device. The ability to become energy self-sufficient and

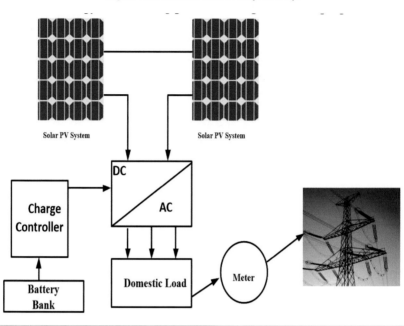

Figure 3.3: Grid tied with battery back-up.

to power remote locations away from the infrastructure are both advantages of using a system that operates independently from the grid. In addition to this, your energy costs are set, and you won't be billed separately for your energy consumption. One more interesting feature of off-grid systems is that they are adaptable, which means that the capacity of the system can be expanded as the user's requirements for energy increase [55], [56].

Because the system is your only source of power, it is common for off-grid systems to have numerous charging sources, such as solar panels, a wind turbine, and a generator. When designing the system, you need to take into account the weather as well as circumstances throughout the year. If snow falls and covers your solar panels, you will need an alternative method to charge your batteries in order to avoid losing power. In addition to this, it is highly recommended that you have a backup generator available in the event that your renewable sources of energy are not able to maintain the batteries charged at all times. One potential drawback is that off-grid devices might not be eligible for participation in certain incentive programs. You also need to design your system so that it can handle 100% of your energy loads and, ideally, even a little bit more than that. Off-grid systems typically consist of more components

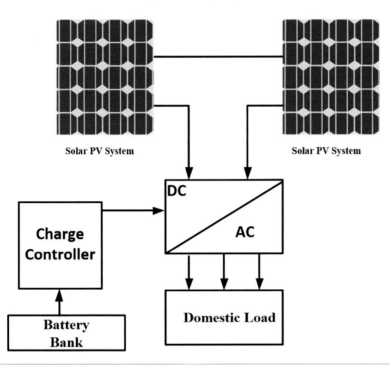

Figure 3.4: Off-grid system.

and come at a higher initial investment than their grid-connected counterparts do. The structure of off-grid system is shown in Figure 3.4.

CHAPTER

4

Solar Charge Controller

4.1 Solar Charge Controller

The pace at which electric current is added to or drawn from electric batteries can be regulated and controlled by a charge controller, also known as a charge regulator, or a battery regulator. It prevents overcharging and has the potential to safeguard against overvoltage, both of which can reduce the performance of the battery or shorten its lifespan and pose a potential threat to safety. Depending on the technology of the battery, it may also conduct controlled discharges or prevent the battery from being completely drained (also known as "deep discharging") in order to preserve the battery's life. The terms "charge controller" and "charge regulator" can apply to either a stand-alone device or to control circuitry that is integrated within a battery pack, battery-powered device, or battery charger. Both of these types of control circuitry can be found in modern lithium-ion batteries. Charge controllers may also include temperature monitoring for the purpose of preventing an overheated battery. In addition to displaying data, transmitting data to distant displays, and logging data, certain charge controller systems also keep track of the electric flow over time [57].

Charge controllers are sometimes referred to as solar moderators when used in conjunction with solar energy systems. Some solar regulators and charge controllers have additional features, such as a low voltage disconnect (LVD), which is a distinct circuit that shuts off the load when the batteries have become overly discharged. A solar charge controller, also known as a regulator, is a small box that is positioned between a solar panel and a battery and is made up of solid-state circuits on a printed circuit board (PCB). They are used to regulate the amount of charge that is coming from the solar panel in order to prevent the battery from getting overcharged. This is accomplished through the use of

29

regulators. In addition to this, it can also be utilized to permit a variety of dc charges while simultaneously supplying the appropriate voltage [58].

When the batteries have reached their capacity, a series charge controller or series regulator will prevent any additional electricity from entering them. When the batteries are at their maximum capacity, the shunt charge controller or the shunt regulator will direct any excess electricity to an auxiliary load known as a "shunt," such as an electric water heater. Charging a battery is disabled by simple charge controllers when the battery's voltage rises above a predetermined threshold, and it is enabled again when the voltage of the battery falls below that threshold. The technologies of pulse width modulation (PWM) and maximum power point tracker (MPPT) are more electronically advanced. These technologies change the charging rates depending on the level of the battery, which enables charging that is closer to the battery's maximum capacity. The system designer is relieved of the responsibility of precisely matching the available PV voltage to the battery voltage when the system contains a charge controller with MPPT functionality. Gains in efficiency of a significant magnitude are possible, particularly when the PV array is situated at a geographically remote location in relation to the battery. A PV array with a voltage of 150 V that is connected to an MPPT charge controller can, for instance, be used to charge a battery with a voltage of either 24 or 48 V. Because a higher array voltage results in a lower array current, the savings in wiring expenses may be sufficient to more than cover the cost of the controller. Charge controllers may also include temperature monitoring for the purpose of preventing an overheated battery. In addition to displaying data, transmitting data to distant displays, and logging data, certain charge controller systems also keep track of the electric flow over time [59].

When used in conjunction with a stand-alone photovoltaic (PV) system, a charge controller's primary purpose is to prevent the battery from being overcharged or discharged to an unsafe level. Any system that has unpredictable loads, user intervention, optimized or undersized battery storage (to minimize initial cost), or any other characteristics that would enable excessive battery overcharging or over discharging is required to have a charge controller and/or low-voltage load disconnect installed. In the absence of a controller, the lifespan of the battery may be reduced, along with the amount of energy that is available. It is possible to create systems to function without the need for a battery charge controller if the load being applied is low, predictable, and constant. It is possible that a charge controller is not required in the PV system if the system designs incorporate oversized battery storage and the battery charging currents are restricted to safe finishing charge rates at an appropriate voltage for the battery technology. Regardless of the system's size or design, periodic shifts in the load profile, or operating temperatures, the correct operation of a charge controller should prevent an overcharge or over-discharge of a battery. The algorithm or control strategy that is utilized by a battery charge controller ultimately plays a role in determining how efficiently batteries are charged, how efficiently PV arrays are utilized, and ultimately how well the system is able to satisfy load demands. The ability of a charge controller to keep a battery in good health, optimize its capacity, and lengthen its lifespan can be improved with the addition of features such as temperature compensation, alarms, and specialized algorithms [60].

A charge controller or charge regulator is essentially a voltage and/or current converter that prevents batteries from being overcharged. Both of these terms refer to the same thing. It is responsible for regulating the voltage and current that are transferred from the solar arrays to the battery. The primary purpose of a charge controller is to ensure that the battery is always in the most fully charged condition that it can be in. The charge controller will prevent the battery from being overcharged and will disconnect the load if it detects that the battery is about to become deeply discharged. In an ideal situation, the state of the battery would be immediately controlled by the charge controller. In between bursts, the controller examines the level of charge currently present in the battery and makes necessary adjustments. The outcome of applying this method is the same as "constant voltage" charging, as the current can be effectively "tapered." If there isn't a charge control, the amount of current that flows into a battery from a photovoltaic (PV) module will be proportionate to the amount of irradiance, regardless of whether or not the battery needs to be charged. If the battery is already at its maximum capacity, then unregulated charging will cause the voltage of the battery to soar to dangerously high levels. This will result in extreme gassing, the loss of electrolyte, increased internal heating, and accelerated grid corrosion.

4.2 Working Principle of a Charge Controller

Pulse width modulation, also known as PWM, and maximum power point monitoring are the two varieties of charge controllers that are utilized in the vast majority of today's solar power systems. Both of these features monitor the temperature of the battery to prevent it from overheating and adjust the charging rate so that it is appropriate for the optimum capacity of the battery. By switching the power devices in the solar system controller, PWM is the most efficient method for achieving consistent voltage during battery charging. When the regulatory mode is set to PWM, the current drawn from the solar array decreases as a function of the state of the battery and the amount of charging that is required. It is the responsibility of the charge controller to guarantee that the system battery is charged and discharged in an effective manner. PWMs are used to help moderate the often-inconsistent voltage that is output by power sources (like solar panels), which helps to prevent the system batteries from being overcharged. When the PWM mode of a solar array is turned on, the charge controller takes over the task of charging the battery in its own special way. It does this by continually monitoring the present state of the battery and automatically adjusting itself to send only the amount of charge that is necessary to the battery. This type of charge controller works by reducing the current from the power source according to the condition of the battery and

Figure 4.1: Pulse width modulation flow.

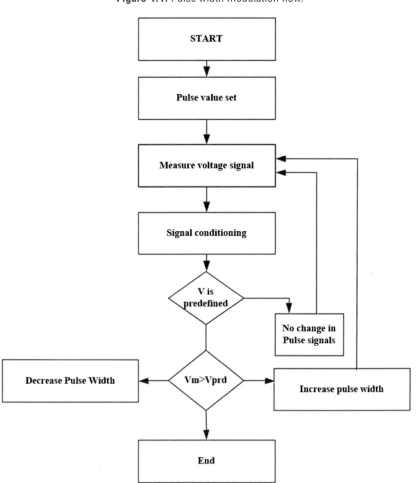

the amount of charging that is required. This is in contrast to on/off charge controllers, which abruptly shut off the power transfer in order to prevent the battery from being overcharged. This is accomplished by the PWM charge controller, which analyzes the current condition of the battery to determine not only how lengthy (wide) the pulses should be, but also how quickly they should be delivered. A flow chart of pulse width modulation is shown in Figure 4.1 [61], [62].

The PWM charge controller will then make the necessary adjustments and send the correct pulse to charge the battery. It will adjust the duration and frequency of the pulses it transmits to the battery based on the requirements of the charging process, as shown Figure 4.1. This is effectively a switch that can be turned on and off very quickly. It is possible that the pulses will be lengthy and continuous when the battery is almost completely depleted; however, as the battery is charged, the pulses will become shorter or will begin to trickle off. This trickle or finish type charging mode is essential for systems that can go days or weeks without using any of their excess energy during times when very little of the solar energy that is available is utilized. This particular type of charge controller is ideally suited for solar arrays that frequently produce more energy than they can use. It offers a number of important advantages, including increased charging efficiency, quick recharging, and healthier batteries that are able to function at their maximum capacity.

4.3 Benefits of a PWM Charge Controller

The traditional systems for charging solar system batteries depended on on–off regulators to limit battery outgassing during periods of excess energy production. However, this frequently resulted in early battery failures and increased load disconnects. The charge controller can gradually decrease the amount of current being supplied to the battery by using a PWM algorithm. This helps to avoid issues with the battery such as gassing and overheating. There are numerous advantages to using a system like this: a PWM charge controller will improve the charging effectiveness, make it possible to recharge the battery quickly, and ensure that the battery continues to have a long and healthy life. In general, a PWM charge controller will provide you with the following benefits: A PWM charge controller will be able to overcome the traditional issues with charge acceptance that are seen in older batteries because it will automatically adjust itself to the requirements of the battery. A PWM prevents problems with gassing and burning, both of which are detrimental to the battery, because it recharges at a faster rate than other charge controllers. The ability to receive a charge is essential for solar system batteries, despite the fact that this has traditionally been a challenge for solar arrays. However, a PWM algorithm can improve the battery's capacity to receive a charge, allowing for a greater amount of the energy generated by the array to be harvested. Equalization is a feature found on a large number of PWM charge controllers. This feature evens out the rate at which a battery cell accepts charge, which helps prevent the capacity from decreasing over time. It is important to keep the state-of-charge as high as possible in order to keep the system in good health and to extend the battery's lifespan. PWM algorithms allow for a greater number of charge and discharge cycles, which results in improved capacity preservation for the battery. Because prolonged undercharging causes grid corrosion and the formation of sulfate crystals on the positive battery plates, sulfation of lead-acid batteries in solar systems is a significant problem that has arisen as a consequence of prolonged undercharging. It has been demonstrated that PWM charge controllers can restore capacity that has been lost over time. This is accomplished by preventing the development of sulphate deposits and overcoming corrosion at the interface. Charge controllers that are older are more susceptible to temperature effects and voltage drops, both of which can have a negative impact and cause problems with the battery's ultimate charge. However, a PWM charge controller will gradually decrease the charge in order to reduce the severity of these effects.

When viewed as a whole, these PWM benefits can be very appealing to PV owners who are looking for an easier way to control the charge that their solar panels produce. However, it is important to be aware of one of the PWM charge controller's drawbacks. The rapid pulses that are produced by the controller have the potential to cause interference for the proprietor when they are operating radios and televisions. Pulse width modulation charge controllers have another drawback, which is that they impose restrictions on the expansion potential of the system. PWM solar adapters employ technology that is analogous to that used in other contemporary high-quality battery chargers. The PWM algorithm gradually lowers the charging current when the voltage of a battery approaches the regulation setpoint. This is done to prevent the battery from overheating and gassing while the charging process continues to deliver the maximum amount of energy to the battery in the shortest amount of time. As a consequence, the charging effectiveness is improved, the battery can be recharged quickly, and it remains healthy even when it is operating at its maximum capacity. In addition to that, the PWM pulsing that this new method of solar battery charging uses promises to bring some very fascinating and one-of-a-kind benefits to the table. Among these are the following:

- The capability to restore lost battery capacity and to disulfate a battery.
- Make significant improvements to the battery's capacity to receive a charge.
- Keep the average battery capacity as high as possible, preferably between 90% and 95%, in comparison to the on–off regulated state-of-charge levels, which are generally between 55% and 60%.

- Bring out-of-balance battery cells back into balance.
- Decrease the amount of burning and gassing the battery produces.
- Make automatic adjustments to account for the deterioration of the battery.
- Solar systems should be able to automatically self-regulate for voltage drops and temperature impacts.

By adjusting the duty ratio of the switches, a PWM charge controller is the most effective means to accomplish constant voltage battery charging. (MOSFET). The amount of electricity drawn from the solar panel by the PWM charge controller decreases as a function of the state of the battery and its capacity for being recharged. The PWM algorithm gradually lowers the charging current when the voltage of a battery approaches the regulation set point. This prevents the battery from overheating and gassing while charging continues to return the maximum amount of energy to the battery in the shortest amount of time. It is planned to bring the voltage of the array down until it is comparable to that of the battery.

The following are some of the benefits that the PWM method offers:

- Improved charging effectiveness, extended battery life, and reduced battery overheating.
- Reduces the amount of strain placed on the battery.
- Capability of removing sulphate from a battery.
- PWM controllers are not the same as DC to DC transformers in any way.

The solar screen can be connected to the battery via the PWM controller, which is a switch. When the switch is in its closed position, the voltage of the screen and the battery will be very close to one another. If we assume that the battery was previously discharged, the starting charge voltage will be somewhere around 13 V, and if we also assume that the cabling and controller will each lose 0.5 V of voltage, the panel will be at 13.5 V. The battery will gradually gain more energy, which will result in a gradual increase in voltage. The PWM controller will begin to disconnect and reconnect the panel in order to prevent overcharge once the absorption voltage has been achieved.

4.4 MPPT Charge Controller

Tracking of the maximum power points solar charge controllers that use maximum power point tracking (MPPT) is distinct from more conventional PWM solar charge controllers in two key respects: first, they are more effective and, second, in many instances they offer a greater number of features. Solar charge controllers that use the MPPT method enable your solar panels to function at the power output voltage that is optimal for them, which can improve their efficiency by as much as 30%. Traditional solar charge controllers compromise the performance of one component of your system in order to accommodate another component's requirements. When compared to "shunt controller" and " PWM" technologies, the MPPT technology is a DC–DC converter that possesses a better level of efficiency. Using a charge controller that does not support MPPT is analogous to connecting the battery straight to the solar panel. A conventional charge controller may charge a battery at a voltage determined by the battery itself. The voltage of a battery after it has been fully charged is, by definition, greater than the voltage of a battery after it has been completely depleted. As a consequence of this, the amount of power that is drawn by an empty battery is typically less than that which is taken by a battery that is full. MPPT, or simply PPT, is a method that is utilized frequently with wind turbines and PV solar systems in order to optimize the amount of power that can be extracted under any given set of circumstances. Although solar power is the primary focus of this article, the principle described here is applicable to a wide variety of sources that produce changeable power, such as optical power transmission and thermophotovoltaics [64]–[66].

PV solar systems can be configured in a wide variety of different ways, depending on how they are connected to inverter systems, other external networks, battery banks, or other types of electrical loads. The main issue that maximum power point tracking (MPPT) attempts to solve, however, is the fact that the efficiency of power transfer from the solar cell depends not only on the quantity of sunlight that falls on the solar panels but also on the electrical characteristics of the load. This is true regardless of where the solar power is ultimately used. As the quantity of sunlight changes, the load characteristic that provides the highest power transfer efficiency also changes. Because of this, the efficiency of the system can be optimized whenever the load characteristic changes in order to maintain the power transfer at the highest possible efficiency. The load characteristic in question is referred to as the maximum power point, and the MPPT procedure refers to the act of locating this point and maintaining the load characteristic there. It is possible to create electrical circuits that present arbitrary loads to photovoltaic cells and then convert the voltage, current, or frequency to suit other devices or systems. The MPPT algorithm solves the problem of selecting the most effective load to be presented to the cells in order to extract the maximum amount of usable power from the system. Solar cells have a complicated relationship between temperature and overall resistance, which results in a non-linear output efficiency that can be analyzed based on the I–V curve. This efficiency can be improved by increasing the temperature of the solar cells. The MPPT system takes a reading of the power produced by photovoltaic (PV) cells and then applies the appropriate amount of load, or resistance, in order to achieve maximum output power under any specific set of environmental circumstances. Devices are generally incorporated into an electric power converter system, which is designed to provide various loads, such as power grids, batteries, or motors, with voltage or current conversion, filtering, and regulation capabilities. Solar inverters change the direct current (DC) power into alternating current (AC) power and may include MPPT; these inverters sample the output power (*I–V* curve) from the solar modules and apply the appropriate resistance (load) in order to acquire the maximum amount of power. The power at the MPP, denoted by "Pmpp," is equal to the combination of the MPP voltage, denoted by "Vmpp," and the MPP current, denoted by "Im" (ImPp).The MPPT solar charge controller is currently the most sophisticated solar charge controller that can be purchased. It is also more expensive due to its increased level of sophistication. In comparison to the PWM charge controller, it offers a number of distinct benefits. At lower temperatures, it has a performance improvement of 30–40%. A synchronous buck converter circuit is at the heart of the maximum power point tracker. The greater voltage from the solar panel is reduced until it is equal to the charging voltage of the battery. It will make adjustments to its input voltage in order to extract the greatest amount of power from the solar panel, and it will then transform this power in order to supply the variable voltage requirements of the load in addition to the battery. It is commonly believed that MPPT will perform better than PWM in environments characterized by low temperatures, whereas the performance of both controllers will be relatively equivalent in environments characterized by temperatures ranging from subtropical to equatorial. The maximum power point tracking charge controller is a direct current to direct current transformer that has the ability to transform power from a greater voltage to power at a lower voltage. Because the amount of power does not change, the product $P = VI$ will continue to have the same value even if the output voltage is lower than the input voltage. This is because the output current will be greater than the input current when the output voltage is lower than the input voltage. Therefore, in order to get the most out of a solar panel, a charge controller should have the ability to select the optimal current–voltage point on the current–voltage graph. This point is known as the maximum power point. This is precisely what an MPPT does. The voltage of the battery that is connected to the output of a PWM controller should, in theory, be the same as the voltage of the controller's input. Therefore, the solar panel is not typically utilized at its maximum power point in the majority of instances [67].

The primary goal of the MPPT system is to draw the maximum amount of available power from PV modules by ensuring that they function at the voltage that maximizes their efficiency. That is to say: the MPPT checks the output of the PV module, compares it to the voltage of the battery, and then determines what the best power that the PV module can generate to charge the battery. It then converts that power to the best voltage to get the maximum current into the battery. In addition to this, it is able to deliver power to a DC load that is directly connected to the battery.

During periods of cold weather, overcast or hazy skies: In general, PV modules function more effectively when temperatures are lower, and MPPT is utilized to extract the greatest amount of power that is available from them. When the battery has been discharged to a significant degree, the MPPT has the ability to draw a greater current and charge the battery, even if the level of charge in the battery has dropped. A charge controller that is embedded with an MPPT algorithm in order to optimize the amount of current that is going into the battery from PV modules is referred to as an MPPT solar charge controller. The maximum power point tracking (MPPT) is a type of DC to DC converter that works by receiving DC input from a photovoltaic (PV) module, transforming that DC to AC, and then converting the AC back to a different DC voltage and current in order to precisely match the PV module to the battery. A boost converter is a type of power converter in which the DC input voltage is lower than the DC output voltage. These are both examples of DC to DC converters. This indicates that the voltage coming from the PV input is lower than the voltage coming from the batteries in the system. A power converter known as a "Buck" converter is one in which the DC input voltage is higher than the DC output voltage. This indicates the PV input voltage is higher than the voltage of the batteries currently installed in the system [68].

Depending on how the system is designed, the MPPT algorithm can be used for either one of them. Buck converters are helpful in most cases when the voltage of the battery system is either equal to or lower than 48 V. On the other hand, a boost converter is the one that should be selected if the voltage of the battery system is higher than 48 V. Off-grid solar power systems, such as stand-alone solar power systems, solar house systems, solar water pump systems, and other similar systems, can benefit from the use of MPPT solar charge controllers [69].

4.5 Development of MPPTs

In partial shading conditions, the incremental conductance technique uses the I–V characteristics to manage the duty cycle of the boost converter, and numerous MPP peaks are split as local MPPs and global MPPs, which are tracked to extract the peak power of the solar PV. The downside of this technique is its low MPP tracking precision, which can be remedied by employing an incremental resistance (INR) technique. At MPP, the derivative of power concerning current is zero in this technique. The INR approach considers various step sizes for tracking MPP. The tracking and convergence speed of MPP increases with step size; however, the oscillations that occur throughout the MPP increase. As a result, a small step size is recommended to reduce oscillations where the convergence speed decreases. However, a tiny oscillation over MPP is addressed by combining a multiple-step size IC with the INR approach.

Using the P&O approach (Figure 4.2), the MPP is monitored in the light of current and historic energy consumption. In the beginning, suitable sensors have been used to detect voltage $V(y)$ and current $I(y)$. After that, a term like $P(y)$ is used to estimate the power. $P(y + 1)$ is an estimate of the current power based on measurements of instantaneous voltage $V(y + 1)$ and current $I(y + 1)$. Next, we evaluate $P(y + 1)$ in light of P [12], the preceding power. There is no adjacent on the duty cycle if the current power is the same as the prior power. After that, make sure the current voltage is higher than the prior voltage, and then the current power is higher than the previous power. If the condition holds, the switching device's duty cycle will be decreased or increased, respectively. However, if the current voltage is lower than the prior voltage, the duty cycle will increase, but in any other case it will decrease. This MPPT method is the simplest one available for monitoring the MPP. The algorithm can follow the MPP in steady-state conditions, but when the PSC happens, it loses the ability to do so [13].

Figure 4.2: P&O flow chart.

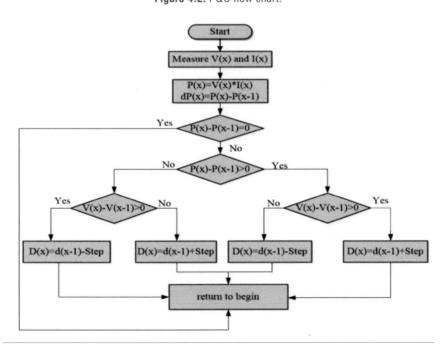

$$\frac{dP}{dV} = I + V\frac{dI}{dV} \tag{4.1}$$

$$\frac{1}{V}\frac{dP}{dV} = \frac{I}{V} + \frac{dI}{dV} \tag{4.2}$$

$$\frac{dP}{dV} = 0 \tag{4.3}$$

At MPP

$$\frac{I}{V} + \frac{dI}{dV} = 0 \tag{4.4}$$

$$-\frac{I}{V} = \frac{\Delta I}{\Delta V} \tag{4.5}$$

The incremental conductance algorithm takes into account both power and voltage changes (Figures 4.3 and 4.4). MPPT functions at its best when the ratio of voltage change to power change is zero. If the ratio of the changes in the PV to the changes in P is greater than zero, then solar photovoltaics (SPV)

Figure 4.3: INC flow chart.

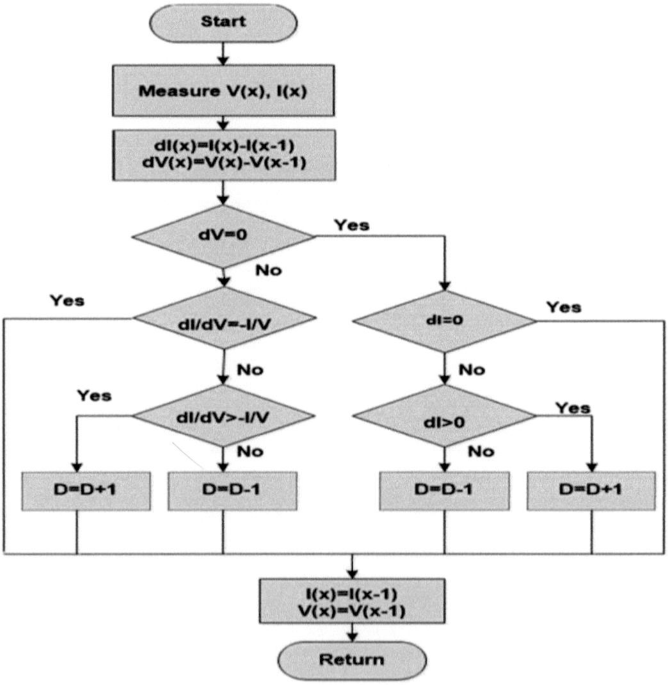

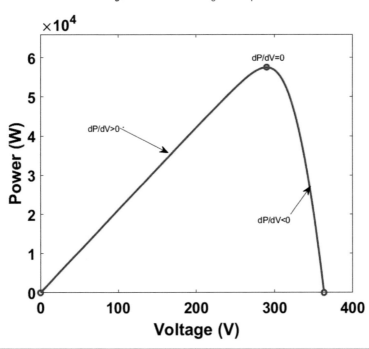

Figure 4.4: INC tracking techniques.

are operating on the left side of the *P–V* curve [14]. Increasing voltage is thus required since more power is being used. As a result, SPV's operational axis is shifting closer to the MPP axis. Since MPP lies on the right side of the *P–V* curve, decreasing the voltage leads to an increase in power, which in turn shifts the SPV's operating point closer to the MPP. Changing the duty cycle can lead to a rise or fall in voltage. The voltage is lowered if the duty cycle is lowered, and raised if the duty cycle is raised. The ANN-based MPPTs are intended to address the shortcomings of the traditional MPPT method. The training phase is important to ANN-based MPPT. Predicting when a peak will occur is a skill that improves with practice [15]. The efficiency of ANN-based MPPT suffers under varying environmental conditions. Therefore, temperature, T, and gradient, G, are being used as input variables in the training of the neural network. The V_{mpp} can be predicted with the use of three-layer neural networks. There are three layers: the input, the concealed, and the output. It is discovered that there is a bias in each successive layer. The V_{mpp} is generated by the output layer after the input has been received by the input layer, processed by the hidden layer, and then output. Error back propagation was utilized to train the neural network [16].

The ANN-based MPPT method gets over the restrictions of conventional MPPT. However, an adequate training data set is needed for ANN-based MPPT in order to train the neural network to predict the correct MPP. If there is a problem with the data, however, a crucial mathematical approach is not necessary to address the non-linearity using fuzzy logic. Fuzzification, a rule base, and defuzzification form the backbone of fuzzy logic controllers. The duty cycle is the output of a fuzzy logic controller, which takes in error signals and fluctuating errors as inputs. The membership function determines how the numbers should be transformed into words. Positive high, positive low, zero, and negative small and negative huge make up the linguistic variables. Thus, it employs five distinct membership functions.

The triangle's membership function is the error plus the error rate of change. The fuzzy inference engine uses a predefined set of rules to generate new rules in response to new errors or shifts in the nature of existing errors. The algorithm produces 25 rules in all. The input variables are "mapped" to the rules. In order to convert a linguistic value into a crisp value, the defuzzification technique is used. After the reference voltage has been filtered to remove unwanted signals, it is sent to the normalization unit, where the input voltage is brought up to specification. The SPV voltage is then compared to a reference voltage V_{pv}, in this case 30 V, to produce an error signal. The duty cycle is regulated by the fuzzy logic controller according to the error and the fluctuating error. The FSC current MPPT approach is a straightforward way to track the MPP of solar PV. The disadvantages of this approach are low MPP tracking precision and excessive oscillations. To address the shortcomings of the general FSC current technique, a hybrid MPPT is developed that employs a mix of FSC current and P&O. The FSC current technique is employed first to track MPP, followed by the P&O technique to decrease oscillations across the MPP. However, due to the inaccuracy of MPP tracking, these solutions are only viable for low-power applications. To track the MPP under varied irradiation conditions, the majority of evolutionary-based MPPT algorithms scan the multidimensional I–V and P–V properties of the PV. Differential evaluation (DE) is a popular MPPT technique that is based on a scanning approach. Multiple iterations are required in DE to determine the MPP position on the P–V curve. As a result, the PV MPP acquired will not be at the intended position. In PV systems, a modified DE MPPT technique is utilized to overcome this issue, and it requires optimum duty cycle information to modify the duty of the boost converter. Furthermore, it provides excellent accuracy in MPP tracking because it does not employ random values to disturb the PV voltage. As a result, when compared to the basic DE technique, the implementation complexity is lowered. The first is DE, which is used to solve non-linear equations. The evolutionary program (EP) is a rarely utilized approach for extracting the greatest output from solar PV. Hashim et al. (2019) presented evolutionary programming to address the shortcomings of artificial intelligence MPPT approaches. Many industrial nonlinear problem-solving applications use the EP approach. It is particularly useful in global MPP tracking of PV under partial shade situations. The ideal value of the multidimensional problem is determined in the EP search space by applying numerous iterations to the development of all individual population solutions. The population generation in each iteration is dependent on the preceding population solution. The mutation factor is employed in population reproduction. During the computational process, the fitness function of each candidate solution is determined. The best fitness function is evaluated among all fitness functions in the computational process for the following iteration. The EP-based MPPT tracking occurs in three steps. In the first stage, various duty values are randomly assigned to all candidates, and in the second stage, the mutation is employed to arbitrarily disrupt the population. Each contender fights against another candidate in the final stage.

Most ANN-based MPPT approaches do not involve any mathematical computation and have a good noise rejection capability. The ANN is equipped with a GA-based MPPT controller to eliminate the DC–DC converter's switching losses. An optimal ANN structure is necessary in the design of the offline neural network in a microcontroller to reduce the size of the MPPT controller. Initially, a genetic optimization technique is utilized to calculate the amount of neurons in the perceptron multilayer network, while ANN is employed to improve tracking speed and eliminate oscillations across the MPP. A PI controller is also used to modify the step size in the GA MPPT approach.

4.6 Limitation of Traditional MPPT

Conventional maximum power point trackers (MPPTs) typically consist of intricate control algorithms and circuitry in order to precisely track the maximum power point of the solar panel. Because of the system's increased complexity, extra components and more sophisticated control methods are required. This might lead to an increase in the system's overall cost. In addition, the complexity of the system makes it more prone to failure or malfunction due to the fact that any problems with the control circuitry or algorithms can interfere with the functioning of the MPPT.

While MPPTs are designed to optimize the energy harvesting efficiency under specified conditions, such as a constant temperature and sun irradiation, real-world scenarios frequently present variable conditions, which might limit the performance of the device under less-than-ideal settings. The output of solar panels can be considerably affected by a variety of environmental factors, including temperature swings, cloud cover, and partial shade. Conventional MPPTs may have difficulty fast adapting to these changing conditions, which can result in inefficient power tracking and a reduction in the efficiency with which energy is converted. Conventional MPPTs have a slow tracking speed, which means that they are unable to modify the operating point of the solar panel in a timely manner in reaction to shifts in irradiance or temperature. If the tracking speed is low, the MPPT might take a longer amount of time to reach the maximum power point, which would result in a lower amount of energy being captured. Because the MPPT might not be able to efficiently keep up with the changes brought on by the rapidly shifting power output of the solar panels, this problem could become even worse.

Because MPPTs need power in order to function properly, the use of these devices can result in additional power losses across the system. The MPPT is responsible for a reduction in the overall energy conversion efficiency of the system due to the energy it consumes. These losses are still present in modern MPPT designs, despite the fact that attempts are made to minimize them as much as possible, and they contribute to a tiny decrease in the overall energy production. The complexity of conventional MPPTs can make them more prone to failures or malfunctions because of their lower reliability and maintenance. It can be difficult to troubleshoot and diagnose problems in the control circuits or algorithms, and doing so may need specialized knowledge or equipment. In addition, the electronic components that are utilized in MPPTs are susceptible to environmental conditions such as moisture, dust, and temperature fluctuations, all of which can have an effect on their reliability over the long run. It is possible that routine maintenance and inspections at regular intervals are required to guarantee correct operation. Cost is a factor to consider, as conventional MPPTs might be more expensive than charge control systems that are less complex. The increased manufacturing and implementation costs are a direct result of the increased complexity of the system, as well as the advanced control algorithms and specialized components. Because of the cost, their use may be restricted, which is especially problematic in contexts where there are financial limitations or when cost-effectiveness is of the utmost importance.

4.7 Merits of Conventional MPPT

The major goal of MPPTs is to improve the energy harvesting efficiency of solar panels. This can be accomplished by increasing the amount of power that can be extracted from the sun. MPPTs ensure that the system collects the maximum amount of energy possible from the sunshine by continuously tracking and adjusting the operating point of the panels to the maximum power point (which is the point at which the panel's output power is maximized). This optimization results in a rise in energy production as well as an improvement in the overall performance of the system.

Conventional MPPTs are designed to function with a wide variety of solar panel technologies, including crystalline silicon, thin-film, and other emerging varieties. This feature allows conventional MPPTs to be more cost-effective. They are able to accept the variable voltage and current characteristics of the many different types of panels, which allows for flexibility in the design and integration of the system. Micropower photovoltaic power converters (MPPT) come pre-programmed with algorithms and control mechanisms that give them the ability to adjust to varying environmental conditions. They are able to immediately react to changes in the amount of solar irradiation, temperature, and shading in order to continuously track the point of maximum power output. Because of its versatility, the system is able to function well in a wide variety of environmental conditions and at all hours of the day. The amount of power that solar panels produce overall can be drastically cut down when they are partially shaded. Conventional MPPTs make use of more complex algorithms, which allow them to dynamically redistribute the flow of power across the panels in order to reduce the negative effects of shade. MPPTs have the ability to ensure that shaded areas do not affect the performance of the entire system by continuously altering the operating point of each panel. This allows for the maximum amount of energy generation possible.

MPPTs are an essential component of solar power systems that incorporate battery storage since they are responsible for the efficient charging of the batteries. They do this by monitoring the voltage of the batteries and adjusting the working point of the panel so that it corresponds to the charging requirements of the batteries. This ensures that the batteries receive the optimal charging current. This feature helps to extend the life of the battery bank and make the most of the useable energy that is stored within it. Monitoring and data logging are capabilities that are offered by a good number of typical MPPTs. They offer information in real time regarding the voltage, current, and power output of solar panels, in addition to other performance characteristics. This data can be put to use for analyzing the system, improving its performance, and identifying and fixing problems. In addition, users are able to monitor the production of energy as well as the operation of the system throughout the course of time thanks to the monitoring feature. Conventional MPPTs have been used for a considerable amount of time and, as a result, they have evolved into a technology that is both standardized and extensively adopted in the solar industry. They have a track record of being reliable and compatible with a wide range of system components and configurations. Because standard MPPTs are so widely available, it is much simpler to track out appropriate items, obtain appropriate levels of technical assistance, and incorporate those devices into solar power systems.

4.8 Bi-inspired MPPT

4.8.1 Particle optimization

The term "swarm intelligence" refers to the coordinated actions of autonomous, distributed systems that take cues from the behavior of living organisms. An iterative optimization technique, particle swarm optimization (PSO) seeks to enhance a particle-representing candidate solution with respect to a quality metric. Particles like these use basic mathematical formulas to navigate the search space by utilizing their position and velocity. Particles are led by the best places in their neighborhood as well as their own best position in the area of search.

The algorithm then sends the duty cycles to the power converter to begin the optimization process, which is represented by a solution vector of duty cycles. In the first iteration, the particles are represented by these duty cycles. Each particle tends to move towards its optimal local position, Pbest. One of these particles is the best in the world, or Gbest. It's the finest way to get a good workout. The duty cycle is shifted to a new location once the velocity is computed, which acts as a perturbation.

4.8.2 Ant colony optimization

Ant colony optimization (ACO) mimics the foraging behavior of ant colonies to find the optimal path to food. The initial concept has since been expanded to address a broader category of numerical issues [66]. ACO is typically used in situations when there are frequent and significant changes to the problem. One of ACO's greatest strengths is that it can operate indefinitely while still adjusting to new information as it comes in. The PV power is used as the target function, and the duty cycle is the regulating variable, in MPPT. Each ant is initially formed in a different starting area. The level of an area's appeal to ants varies from place to place. The ants shift from lower to higher attraction strength as a result of the disparity. The chance of an ant's movement is determined by the density of the surrounding area. In the next version, the ants will advance to a more powerful stance. The relative attractiveness of each area is once again determined. The ants' iterative process leads them closer and closer to the MPP, or maximum population point.

4.8.3 Genetic algorithm (GA)

The evolutionary algorithm class includes GA. To be more precise, it is an approach to problem solving that draws inspiration from evolutionary theory. The procedure involves categorizing some inputs as chromosomes, which are then subjected to recombination and mutation in order to see if they satisfy a specified fitness function. Since the goal of evolution is to produce a more robust species, GA discovers the optimal solution by mixing and matching genes at random.

The optimization problem's starting chromosomal set is the search parameters. Both voltage and duty cycle can be used as such factors in MPPT. The PV equation is the fitness function, whereas real or binary integers can be used to define chromosomes. Because larger populations converge more quickly, but generate more data at once, choosing a chromosomal length is crucial. The chromosomes' DNA is then modified by the crossover and mutation operation, resulting in a new generation of chromosomes. A new fitness value is calculated based on how well this new generation performs in the fitness function. The MPP optimization is performed by repeatedly iterating over the parameters and selecting the chromosome with the highest fitness value.

4.8.4 Grey wolf optimization

Grey wolf optimization (GWO) is a meta-heuristic strategy heavily influenced by the optimization of the grey wolf's attacking strategy during hunting. This method can successfully mimic the social structure and hunting prowess of grey wolves. Alpha (), Beta (), Delta (), and Omega () grey wolves are used to accurately model different ranks in the pack's hierarchy. The optimal answer for this biomimetic method is supposed to be in the corresponding mathematical model.

Then, the two remaining candidates are ranked as the second and third best answers, respectively. Hunting, pursuing, and tracking prey are the first two steps in GWO, followed by creating a group, encircling the prey, and eventually assaulting it. When developing the GWO to solve optimization issues in MPPT for PV modules, this global hunting mechanism is incorporated into the design process. In the grey wolf pack, the clans act as leaders and are followed by the clans when it comes to hunting strategy. The clan's primary responsibility is to treat all of the injured wolves in the pack. To dampen steady-state oscillations, we are using the GWO described in [10] in conjunction with direct control of duty cycle to maintain a constant duty cycle at the maximum power point (MPP).

4.8.5 Ant bee colony algorithm

Simple, with few regulated parameters and independent algorithm convergence criteria from initial conditions, artificial bee colony (ABC) algorithms are the primary bio-inspired approach mentioned in [15,16]. Multidimensional and multimodal optimization issues are simple for this swarm-based meta-heuristic approach to solve. The artificial bees can be broken down into three distinct types: workers, observers, and scouts. The worker bees are classified as either employed (those actively searching for food or exploiting a food source) or onlookers (those waiting in the hive for the queen to make decisions about which food source to use) and the scout bees (those carrying out the random search) are employed (those carrying out the search). The ideal answer is reached with less time investment thanks to the coordinated efforts of all three groups. Here, the duty cycle represents the feeding position, and peak power output serves as the algorithm's fuel.

4.9 Characteristics of an MPPT Solar Charge Controller

In any applications in which a PV module serves as the energy source, an MPPT solar charge controller is utilized to make the necessary adjustments for identifying variations in the current–voltage characteristics of a solar cell, which are represented by an *I–V* curve.

An MPPT solar charge controller is essential for any solar power system that needs to extract the maximum amount of power from PV modules. This type of charge controller requires PV modules to operate at voltages that are relatively close to their maximum power points in order to draw the maximum amount of available power. Users are able to use PV modules with a greater voltage output thanks to the MPPT solar charge controller. This is in contrast to the operating voltage of the battery system. For instance, if a photovoltaic (PV) module needs to be positioned at a great distance from the charge controller and the battery, the module's wire size needs to be increased significantly to minimize voltage loss. Users are able to wire photovoltaic (PV) modules for either 24 or 48 V (depending on the charge controller and PV modules) and bring electricity into 12 or 24 V battery systems when utilizing an MPPT solar charge controller. This results in a reduction in the required cable size while maintaining the maximum output of the PV module. The MPPT solar charge controller can decrease the overall complexity of the system while maintaining a high level of system efficiency. In addition, it is adaptable for use with a wider variety of energy sources. Because the output electricity from the PV system is used to directly control the DC–DC converter. The solar PV with MPPT is as shown in Figure 4.5.

It is possible to use an MPPT solar charge controller with other types of sustainable energy sources, such as small water turbines, wind-power turbines, and so on.

Figure 4.5: MPPT charge controller.

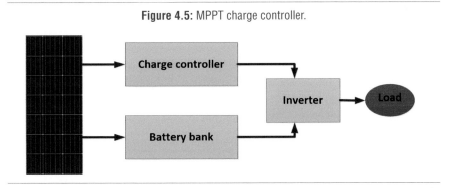

4.10 MPPT with a PWM Charge Controller

If maximizing charging capacity was the only factor that was taken into consideration when designating a charge controller, then everyone would use an MPPT controller. However, the two technologies are distinct and each has its own set of benefits. The choice is dependent on the conditions of the location, the components of the system, the size of the array, the load, and the cost of a specific solar panel system. The following is a comparison of them. An MPPT controller is recommended for use in environments with lower temperatures, and is able to collect the additional electricity from the module and use it to charge the batteries. In comparison to a PWM controller, it can generate up to 20–25% more charging. Because the PWM technology charges at the same voltage as the battery, the PWM type is unable to collect any excess voltage that may be present. When solar panels are installed in warm or hot climates, however, there is no surplus voltage to be transferred. This renders the MPPT superfluous and eliminates any advantage it may have had over a PWM. For PWM, the voltages of the PV array and the battery should be the same; however, for MPPT, the PV array voltage can be greater than the battery voltage. PWM operates at the battery voltage, which is why it functions well in warm temperatures and when the battery is almost at its maximum capacity. On the other hand, MPPT operates at a voltage higher than the battery voltage, which is why it can provide "boost" in cold temperatures and when the battery is low. PWM is generally recommended for use in smaller systems because the benefits of MPPT are not as significant there. On the other hand, MPPT is recommended for use in systems that are 150–200 W or greater in order to make the most of its benefits. MPPT controllers are generally more expensive than PWM controllers, but because they are more efficient under certain conditions, they can generate more power with the same number of solar modules than a PWM control can. This is because MPPT controllers have a higher power point tracking (MPPT) capability [70].

The majority of high-quality charge controller devices have what is referred to as a three-stage charge cycle, which consists of the following stages. At this point, the battery will be able to take in all of the electricity that is being supplied by the solar array. The magnitude of this current will have the same value as the short circuit current I_{sc} that is generated by the solar array. During the bulk portion of the charge cycle, the voltage gradually increases to the bulk level, which is typically between 14.4 and 14.6 V, while the batteries draw the maximum amount of current possible. The absorption stage will commence once the bulk voltage level has been reached.

During this phase, the voltage is kept constant (maintained at bulk voltage level) for a specified amount of time (usually an hour), while the current progressively decreases as the batteries charge up. This happens while the voltage is maintained at bulk voltage level. This is done to prevent the battery from overheating and over gassing itself. As the battery is charged to its maximum capacity, the current will gradually decrease to amounts that are safe.

When a battery reaches its maximum capacity, switching to the float stage will provide a very low rate of maintenance charging while simultaneously decreasing the amount of heating and gassing that a fully charged battery experiences. When the battery has been completely refueled, there is no longer any room for chemical processes, and the entire amount of charging current is converted into heat and gassing instead. The function of the drift state is to shield the battery from damage caused by sustained overcharging. Following the completion of the absorption time period, the voltage will drop to the float level. This is generally the case (typically 13.4 to 13.7 V) for a 12 V battery, and the batteries draw a small maintenance current until the next cycle begins. The battery charging and discharge sequence as shown in Figure 4.6.

Figure 4.6: Relationship between the voltage and the current.

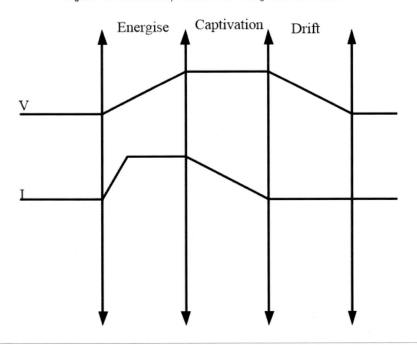

Shunt regulation and series regulation are the two fundamental approaches that can be utilized in order to control or regulate the discharge of a battery. Although these two approaches are both useful, there are numerous permutations that can be made to each approach that can fundamentally change how they work and the situations in which they can be used. The term "series type" refers to the controller that is utilized when the MOSFET switch is connected in series with the PV array and the battery. The term "shunt type" refers to the configuration in which it is connected in parallel across the PV array or the battery. When the battery has reached its maximum capacity, the series type keeps the MOSFET switch in its open position. During this time, the PV array will not be able to deliver any current. In the shunt type, the MOSFET switch is maintained in its closed position so that the entire short circuit current of the PV array can be shunted (diverted) away from the battery when the battery has reached its maximum capacity for charge [71].

4.11 Design of a Series Controller

The primary purpose of the series charge controller is to regulate the voltage and current flow between the solar panel and the battery. Additionally, it can be used to prevent the batteries from being overcharged and to disable current flow when the batteries have reached their maximum capacity for charging, both of which contribute to an improvement in the battery's overall performance. This form of controller operates in series between the array and the battery, as opposed to parallel operation like the shunt controller, which operates between the array and the battery. There are a few distinct iterations of the controller known as the series type; however, they all share the characteristic of placing some kind of control or regulatory element in series with the battery and the array. Due to the current limitations of shunt controllers, this type of controller is not only the practical choice for larger PV systems, but it is also the type of controller that is frequently used in smaller PV systems. In a design for a series controller, a relay or solid-state switch either opens the circuit between the array and the battery to stop charging or limits the current in a series-linear fashion to keep the battery voltage at a high value. This can either stop charging entirely or allow the charging to continue. In the more straightforward series interrupting design, the controller re-establishes the connection between the array and the battery once the voltage of the battery reaches the set threshold for the array reconnect. Because the battery is gradually becoming completely charged throughout these on–off charge cycles, the amount of time that the device is "on" is gradually decreasing. Because a series controller opens the array rather than short-circuiting it, as shunt controllers do, when the controller modulates, a blocking diode is not required to prevent the battery from short-circuiting. This is because the series controller open-circuits the array [72].

When the batteries have reached their capacity, a series charge controller will prevent any additional current from entering them. The controller depicted in Figure 4.7 is a variety that operates in series with both the array and the battery. There is more than one variation of the controller known as the series type, and each of these variations uses some kind of control or regulatory element in series. Either a relay or solid-state switch can either open the circuit between the array and the battery to stop the charging process or limit the current in a series-linear fashion to keep the voltage of the battery at a high value. This can be accomplished by either opening the circuit. The charge controller configuration is as shown in Figure 4.7.

Figure 4.7: Charge controller for a series.

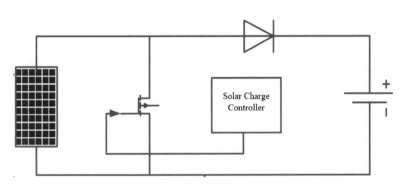

4.12 Design of the Shunt Controller

The shunt charge controller is a form of charge controller that can be utilized to regulate the voltage as well as the current flow that occurs between the load and the battery, as shown in Figure 4.8. In addition, it can be used to redirect the excess charge to the shunt load (for example, an electric water heater), as well as to provide a continuous supply of power to the load at all times without any interruptions. When the batteries are at their full capacity, a shunt charge controller will direct any excess electricity to an auxiliary load, also known as a "shunt," such as an electric water heater. The shunt charge controller is shown in Figure 4.8.

In contrast to batteries, photovoltaic cells are designed to produce a limited amount of current; as a result, PV modules and arrays can safely be short-circuited without causing any damage. Shunt controllers are able to function on the premise of the ability to short-circuit individual modules or an entire array. By performing an internal short circuit on the PV array, the shunt controller is able to control the rate at which a battery is charged using power from the PV array. When the array is in the process of regulating itself, all shunt controllers are required to have a blocking diode connected in series between the battery and the shunt element. This is done to prevent the battery from being short-circuited. Because of factors such as the resistance of the shunt element, the cabling, and the fact that the array is never completely short circuited, there is always some power loss within the controller. This is caused by the fact that there is a voltage drop between the array and the controller. Because of this requirement, most shunt controllers can only be used in photovoltaic systems

Figure 4.8: Shunt charge controller.

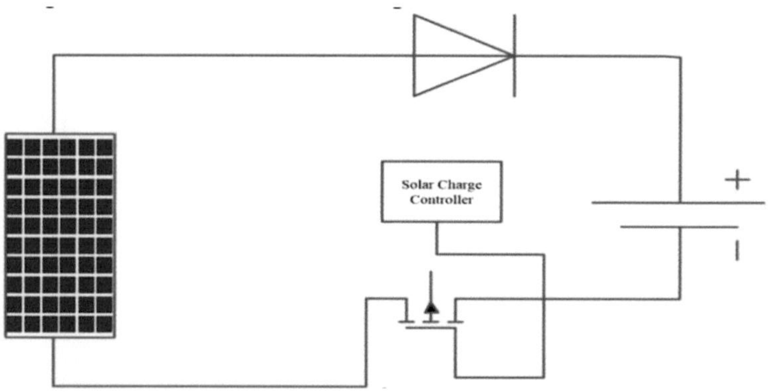

that have array currents of less than 20 A because they also require a heat sink to dissipate power. Depending on the particular design, the regulation element in shunt controllers is most often a power transistor or a MOSFET. However, there are exceptions. There are a few different approaches that can be taken when designing a shunt controller. The first type of controller design is a straightforward interrupting, also known as an on–off, type. The second type is one that restricts the array current in a step-by-step fashion, and it does so by gradually increasing the resistance of the shunt element as the battery reaches its fully charged state [73].

The shunt controller is located inside the charge controller and is responsible for regulating the charging of a battery using the PV array. When the array is in the process of regulating itself, all shunt controllers are required to have a blocking diode connected in series between the battery and the shunt element. This is done to prevent the battery from being short-circuited. Depending on the particular design, the regulation element in shunt controllers is most often a power transistor or a MOSFET. However, there are exceptions. The voltage can range anywhere from 10.5 to 14.4 V in a battery system that operates on 12 voltzite voltage of a typical fully charged and discharged battery ranging between 12.4 and 12.7 V when there is no current flowing to charge or discharge the battery. When loads are turned on, the voltage drops down to a lower level, such as 12 V or 11.8 V, while it jumps up to a higher level, such as 13.7 V (depending on the current) when charging current is flowing through the circuit. This occurs because the voltage is proportional to the current.

The reaction that takes place within a battery when it is being deeply discharged takes place close to the grids and causes the bond between the active materials and the grids to become less strong. When a battery is repeatedly subjected to a deep discharge, both its capacity and its lifespan are eventually reduced. The majority of charge controllers come equipped with an optional feature that will disconnect the system loads once the battery reaches a low voltage or low state of charge condition. This is done to prevent the battery from suffering a deep discharge. In the event that the voltage of the system remains below 11.5 V for a period of at least 20 s, the charge controller will be turned off for a period of at least 30 s. A delay of 30 s is built into the system as a safeguard against any kind of swinging situation [74].

5

Necessity of Boost converters

5.1 Necessity of Boost converters

Boost converters are a specific kind of DC–DC converter that are frequently utilized in photovoltaic (PV) systems to raise the voltage of the DC electricity that is produced by the solar panels. The output voltage of a solar panel is generally low and fluctuates depending on the amount of sunshine that is available. By utilizing a boost converter, the voltage can be increased to a level that is appropriate for charging batteries or activating DC loads. This makes the boost converter an extremely versatile piece of electrical equipment. Boost converters are frequently utilized in solar photovoltaic (PV) devices for the following reasons. In order to transform the low voltage DC power that is produced by the solar panels into a higher voltage that is suitable for charging batteries or powering DC loads, an effective method known as a boost converter is required. A significant number of solar photovoltaic systems are intended to charge batteries, which call for a particular voltage range in order to charge effectively. Through the utilization of a boost converter, the voltage produced by the solar panel can be brought up to the level required for effectively charging the batteries. It is possible to use a boost converter in conjunction with an MPPT controller to ensure that the solar panel is functioning at its highest power capacity. This, in turn, increases the amount of power that can be gleaned from the sun. Boost converters can also be used to accommodate for voltage drops in lengthy cable runs between solar panels and the charge controller or the battery bank. This can be accomplished by connecting the solar panels directly to the boost converter. In general, a boost converter is an essential part of many solar photovoltaic (PV) systems because it permits effective power conversion and makes it possible for the system to function to the fullest extent of its potential [75], [76].

5.2 Components Used in a Boost Converter

The following are the primary components that are employed in a boost converter circuit according to a conventional configuration. This is the DC power source that is being improved, which may be a battery or a photovoltaic system depending on the circumstances. The direct current power supply is the primary source. In most cases, the power switch is a transistor or a MOSFET, and it is responsible for rapidly switching the incoming electricity on and off at a high frequency. This task of transitioning is carried out in a cyclical fashion by it. When the power switch is turned on, energy is stored by an inductor in a magnetic field. This energy is then released when the switch is turned off, releasing all of the stored energy. Inactive components are what make up an inductor. This component, known as a diode, is responsible for preventing the current from travelling in the opposite direction of that intended by the inductor. A capacitor is utilized by the boost converter so that the voltage that is generated by the device can be maintained in a more stable state. The word "load" refers to the appliance or circuit that is receiving its electricity from the higher voltage. Loads can be either direct or indirect. Control circuit: this component, which is typically a microcontroller or an analogue control circuit, is in charge of monitoring the output voltage and modifying the duty cycle of the power switch in order to maintain a continuous level of output voltage. This component is responsible for keeping the output voltage at a consistent level. These components work together in a synchronized fashion to raise the input voltage to a higher output voltage, which is suitable for either operating a load or charging a battery. The higher output voltage is achieved by raising the input voltage. The greater output voltage can be obtained by increasing the voltage that is being supplied [77], [78].

5.3 Necessity of a Buck converter

Buck converters, which are also known as step-down converters, are frequently utilized in the power electronics industry to bring the input voltage down to a more manageable level before being converted into an output voltage. In certain applications, buck converters are required for a variety of purposes, including efficiency in the use of energy. Buck converters can achieve efficiencies of 95% or higher, making them one of the most effective types of power converters. This is due to the fact that their operation involves turning the incoming voltage on and off at a high frequency, which in turn decreases the amount of power that is wasted as heat. Buck converters are frequently used in battery-powered devices to step down the voltage of the battery to a lesser voltage that is appropriate for powering the electronics inside of the device. This allows the battery to more efficiently power the device. This serves to prolong the life of the battery while simultaneously reducing the required size of the battery. Buck converters are a useful tool for regulating the output voltage of a power supply or voltage source. This can be accomplished through the process of voltage modulation. They are able to keep the output voltage constant regardless of whether the input voltage is changing or the load is being applied [79].

Buck converters can be built to withstand high current loads, which qualifies them for use in operating high-current electronics like motors, LEDs, and other high-current gadgets. Buck converters can help reduce noise and interference in electrical circuits by filtering out high-frequency noise and ripple on the incoming voltage. This can help reduce the amount of noise that occurs in the circuit. In general, buck converters are an essential component of a variety of applications involving power electronics. This is especially true for those applications involving battery-powered devices or voltage management. As a result of their high efficiency, small size, and accurate voltage regulation, they are ideally suited for a wide variety of applications.

5.4 Necessity of a Buck Boost Converter

The input voltage can be stepped up (boost) or stepped down (buck) by a buck–boost converter, which is a form of DC–DC converter. The type of step the input voltage takes depends on the requirements of the application. The following is a list of some of the most important functions and benefits that come from utilizing a buck–boost converter. Buck–boost converters can operate over a wide range of input voltages, which makes them useful in applications where the input voltage may change greatly or where numerous input sources are used. This ability to operate over a wide range of input voltages gives buck–boost converters their name. Regulation of voltage buck–boost converters are able to regulate the output voltage of a power supply or voltage source, even when the incoming voltage fluctuates. This is accomplished through a process known as "bucking." Because of this, they are helpful in applications that require a steady output voltage. Buck–boost converters can be used to charge batteries from a wide range of input voltages, making them helpful in applications where the battery voltage may fluctuate greatly. Buck–boost converters can be used in inverter applications, where they can transform a DC input voltage into an AC output voltage with a particular waveform. Buck-boost converters can achieve efficiencies of 95% or higher, making them among the most efficient power converters. Because of this, they are helpful in applications where it is essential to maximize energy efficiency, such as in devices that are powered by batteries. In general, the buck-boost converter is an adaptable and helpful component in power electronics. It enables effective power conversion and voltage regulation over a broad variety of input voltages, which makes it an extremely useful tool [79], [80].

5.5 Effect of Duty Cycle

The term "duty cycle" is used to describe the ratio of the amount of time that the switch is switched on to the overall amount of time that the boost converter is switching. Because it decides the voltage that is produced by the boost converter, the duty cycle is an extremely important component. The inductor in the circuit stores energy when the switch is turned on, and when the switch is turned off, the energy that was stored is discharged into the output capacitor, which results in an increase in output voltage. The amount of energy that is retained in the inductor is directly proportional to the duty cycle, which is the amount of time that the switch remains on while it is operating. If the duty cycle is too low, then the inductor will not be able to retain an adequate amount of energy, which will result in a reduced output voltage. If the frequency of the duty cycle gets too elevated, on the other hand, the inductor might not have enough time to discharge completely, which would result in an unsteady output voltage. As a result, the duty cycle needs to be adjusted so that it corresponds adequately with the intended output voltage. In conclusion, the duty cycle plays an essential part in controlling the output voltage of a boost converter, and it is essential that it be meticulously regulated in order to guarantee that the circuit will operate in a reliable and effective manner. It is necessary to have knowledge of both the operating frequency and the output voltage of a converter in order to make an approximation of the duty cycle of the converter. When referring to a converter, the term "duty cycle" refers to the amount of time – expressed as a percentage – that the converter's switch is "on" during an entire toggling cycle. It is determined by applying an approach that is as follows:

The boost converter voltage level either steps down or up based on the time duration of the on-time and off-time of the switching device used on the circuit topology. The converter's switch-on time is estimated by the following formula:

$$T_{on} = (V_{out}/V_{in})\star(1 - D)$$

where V_{out} is the voltage that is produced by the converter, V_{in} equals the voltage that is supplied to the translator, D equals the decrease in voltage across the switch.

Use the following calculation to determine the amount of time that the switch in the converter will be off:

$$T_{off} = (1 - D)^{\star} \ T_{on} \ /D.$$

You will be able to make an educated guess regarding the duty cycle of a converter if you follow these procedures. It is essential to keep in mind that the duty cycle can change based on the load in addition to other variables; consequently, this assessment might not be accurate in all circumstances.

In order to grasp intricate systems and make well-informed choices, modelling and analysis are indispensable. The reasoning behind the modelling and research is as follows. Future results or behaviors of a system can be predicted using models. This is especially helpful in areas where accurate predictions can help us prepare for the future, such as finance, engineering, and climate science. Systems and procedures can be optimized with the help of models. In manufacturing, for instance, models can help figure out how much raw materials to use to cut expenses without sacrificing quality. Complex processes are often difficult to comprehend without the aid of models. Understanding how a system works as a whole requires deconstructing it into its component components and examining their relationships with one another. You can use a model to check your theory or hypothesis. By modelling a system and analyzing its performance, we can tell if our assumptions hold up or if we need to make changes. Risk in a variety of contexts can be evaluated through modelling and research. This is especially helpful in the insurance and finance industries, where an accurate assessment of risk is essential for making sound choices. To sum up, modelling and analysis are valuable resources for learning about and managing uncertainty in complicated systems and making informed decisions about where to allocate resources. They are crucial for competent decision making across many disciplines.

5.6 Circuit Configuration of a Boost Converter

Output current $(I_o) = \frac{P}{V}$

Current ripples $(\Delta I) = 0.05 * I_o * \frac{V_o}{V_{in}}$

Voltage ripple $(\Delta V) = 0.01 * V_o$

Figure 5.1: Maximum power point tracker.

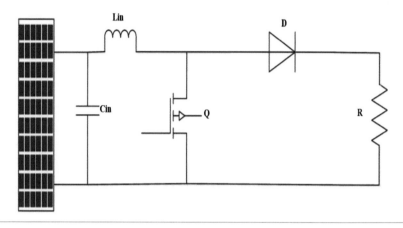

Inductance $L = \frac{V_{in}(V_o - V_{in})}{\Delta I * f_s * V_o}$

Capacitor $C = \frac{I_o(V_o - V_{in})}{f_s * \Delta V_o * V_o}$.

The power output of photovoltaic (PV) solar panels can be maximized with the help of an electronic device known as a maximum power point tracker, or MPPT for short, as shown in Figure 5.1. The MPPT functions by making constant adjustments to the load that is placed on the solar panels in order to keep those panels working at their maximum power point (the point on the voltage–current curve where the panels produce the most power). Due to the power output of a solar panel being extremely sensitive to the operating conditions, such as temperature and irradiance levels, it is essential to make use of a MPPT device. In the absence of an MPPT, the solar panels might work at less-than-ideal points on the voltage–current curve, which would lead to a reduction in the total amount of power produced. An MPPT can, besides improving the electrical output of the solar panels, protect the panels from damage by preventing overloading and overcharging. This is accomplished by preventing the panels from being overloaded. Because of this, the solar panels may have a longer lifespan and require fewer dollars in upkeep over their lifetime. When it comes to maximizing the effectiveness and output of PV solar panels in a variety of applications, including residential, commercial, and industrial settings, it is absolutely necessary to make use of a maximum power point tracker (MPPT).The topology of a boost converter as shown in Figure 5.1.

Track the MPP using the present power and the power from the preceding measurement in the perturb and observation technique. In the first step of the process, voltage $V(x)$ and current $I(x)$ were determined with the help of the appropriate instruments, as shown in Figure 5.2. The power is then approximated using a notation known as $P(x)$. After some time had passed, measurements of instantaneous voltage $V(x + 1)$ and current $I(x + 1)$ was taken in order to derive an estimate of the current power, which was denoted by the symbol $P(x + 1)$. After that, the current power, $P(x + 1)$, is contrasted with the power that was present before. There is no adjacent on the duty cycle if there is no difference between the power that is being produced now

Figure 5.2: Flow chart of MPPT.

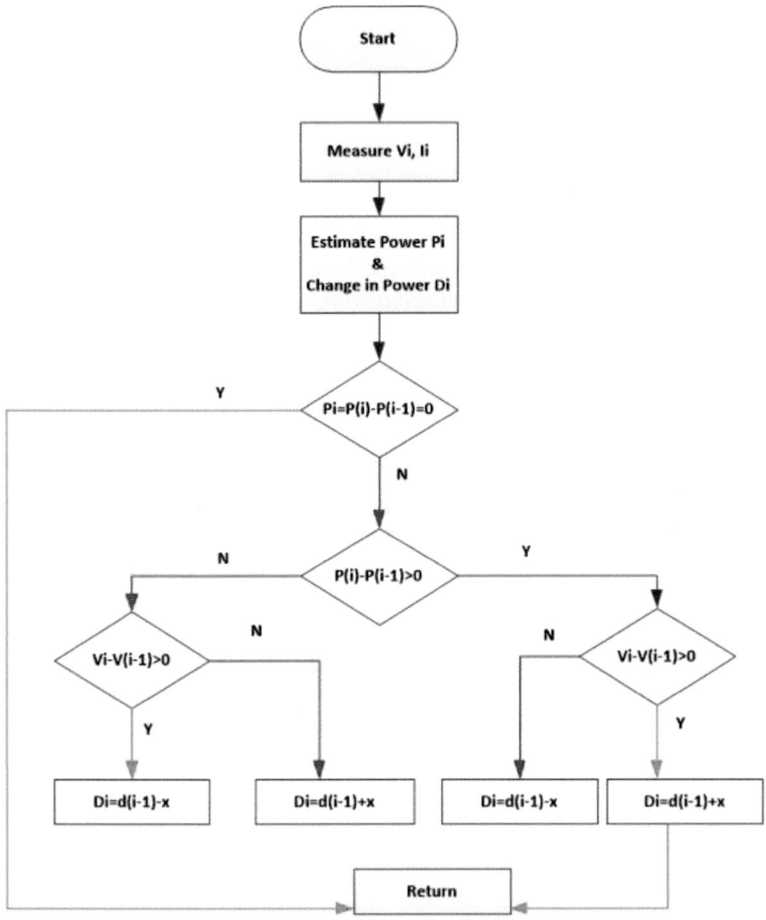

and the power that was being produced before. Figure 5.2 presents P&O's organizational structure in the form of a flow diagram. Check again later to see if the differential between the present power and the previous power is greater than zero, and then check again later to see if the present voltage and the previous voltage are both higher than zero. If the requirement is met, either the switching device's duty cycle should be decreased or it should have an increased duty cycle. If, however, the difference in voltage between the current voltage and the preceding voltage is greater than zero and the requirement is satisfied at this instant, the duty cycle will be increased; otherwise, it will be decreased. It is the most basic MPPT algorithm possible to use for tracking the MPP. Under steady-state conditions, the algorithm is able to successfully track the MPP; however, if a situation that involves a PSC or rapid climate change happens, the algorithm is unable to successfully track the MPP [82]–[85].

5.7 Result and Analysis

The relationship between voltage and time is shown in Figure 5.3. Initially an overshoot occurs, and after a few seconds it will reach to a stable position. The conventional boost converter with conventional MPPT is taken to settle to the stable value. The output RMS voltage of the boost converter is 311.2 V. The wave

Figure 5.3: Output voltage vs. time.

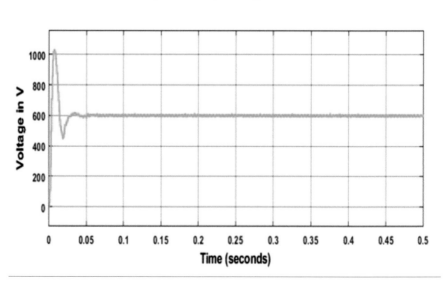

form pattern is similar to the pattern of voltage. The RMS value of the out power is 4999 W. Hence the proposed converter efficiency is higher than that of others. The output power in shown in Figure 5.4.

Figure 5.4: Output power vs. time in seconds.

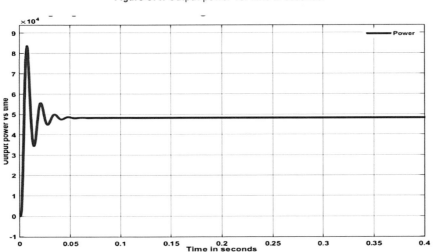

5.8 Modified Boost Converter

The suggested module contains 47 strings that are connected in parallel and 10 strings that are connected in series. The photovoltaic system has a maximum power output of 249.9 W, the open-circuit voltage is 43 V, there are 60 cells in each module, the short-circuit current is 7.75 A, and the necessary current at maximum power is 7.1 V. The boost converter receives this incoming power through its feed. At first, the simulation was run at a power of 1000 W and a temperature of 25o. Figure 5.5 depicts the improved boost converter. As can be seen in Figures 5.6 and 5.7, the converter's output power varies due to the non-steady relationship between output current and time.

The simulation is carried out with continuous irradiations, despite the fact that the solar photovoltaic (PV) system does not generate a consistent output power as demonstrated in Figures 5.6 and 5.7. It is true that the voltage and the electricity both generate the same result, and it is also true that the opposite

Figure 5.5: Modified boost converter.

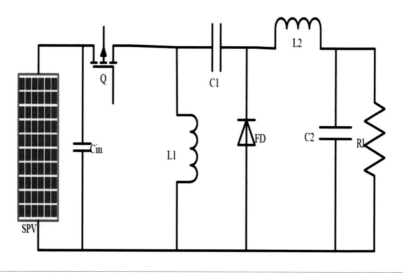

Figure 5.6: Current vs. time.

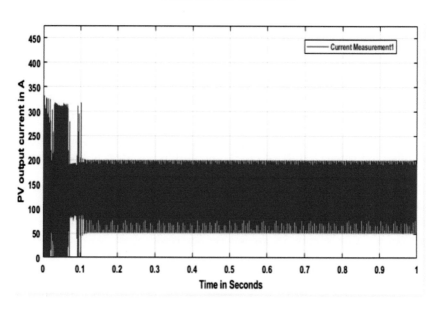

Figure 5.7: Power vs. time.

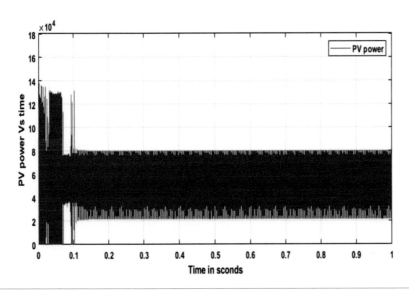

is true. However, with the help of the coupling capacitor, the outbound power of the solar photovoltaic system can be provided to the boost converter. The task of the coupling capacitor is to balance out the current fluctuations that are produced by the solar input. These fluctuations are created by the sun as shown in Figure 5.8. An enhancement in the voltage profile is brought about by making use of the input inductors, and this is done on the basis of the pulse data that is received by the switch. There are no ripples in the out current, as demonstrated by the fact that it took significantly less than 0.025 s to achieve a steady number. Additionally, there are no overtones present in the output power, which is 4×10^4 W. It illustrates the relationship between output power and duration, as can be seen in Figure 5.9. The simulation is run using a changing irradiation codec. The output power of the proposed model is 4.525×10^4 W when measured at 1000 W/m^2, but this value drops to 3.56 W when measured at 700 W/m^2, and it reaches 3.45 W when measured at 500 W/m^2, as shown in Figures 5.10 and 5.11. Based on the calculation, it was discovered that the fluctuation of variable irradiations has a significant impact on the amount of solar photovoltaic energy produced. The proposed circuit configuration includes a maximum power point detector, which makes it capable of responding appropriately to variations in irradiations.

Figure 5.8: Output current vs. time.

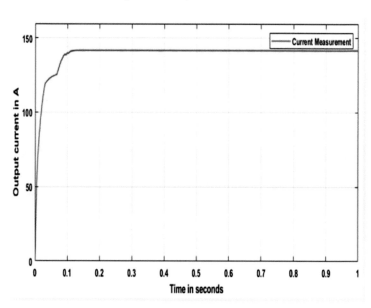

Figure 5.9: Output power vs. time.

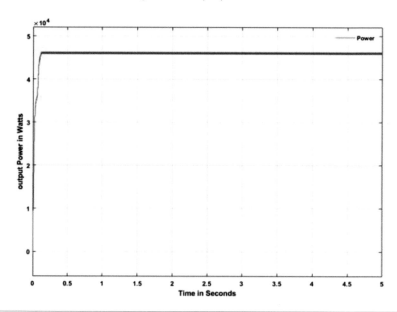

Figure 5.10: Irradiation variations.

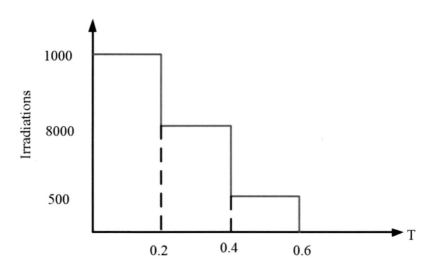

Figure 5.11: Converter under variable irradiations.

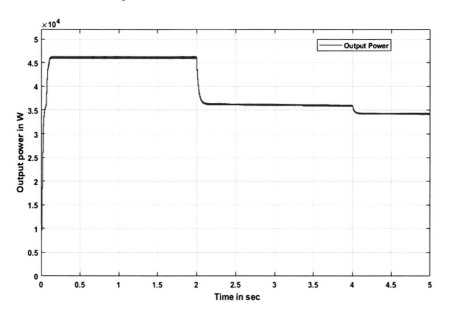

6

Necessity of a Hybrid Power Plant

6.1 Necessity of a Hybrid Power Plant

Combined-cycle power plants, also known as hybrid power plants, are a type of power production facility that uses more than one source of energy to generate electricity. The sources of electricity can be conventional fossil fuel-based sources like coal or natural gas, as well as sustainable energy sources like wind, solar, hydro, or geothermal power. Because no single energy source can consistently and efficiently satisfy all of our requirements for energy, we are forced to rely on hybrid power plants to meet our power generation requirements. It is important to note that renewable energy sources, like wind and solar, do not generate electricity continuously and their production can fluctuate depending on the conditions of the surrounding environment, including the weather. Even though they are a dependable source of energy, sources that rely on fossil fuels are detrimental to the ecosystem and contribute to climate change. Hybrid power plants are able to supply a source of electricity that is more consistent and dependable because they combine a variety of different energy sources. For instance, a combination power plant that relies on solar and wind energy to generate electricity can continue to do so even when one of the sources is unavailable. In addition, hybrid power facilities have the potential to cut pollution of greenhouse gases and to encourage equitable growth. It is also possible for hybrid power plants to increase energy security while simultaneously decreasing reliance on a single type of energy source. Hybrid power plants are able to help countries become less dependent on the purchase of fossil fuels and provide more energy independence by diversifying the energy blend in power plants [86], [87].

6.2 Grid Connected Solar PV system

Figure 6.1: Three phase grid connected solar PV.

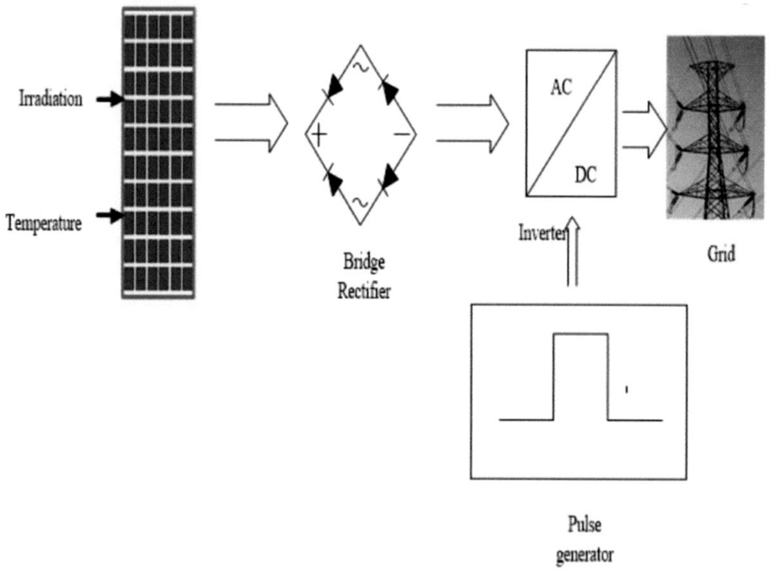

The solar system that is connected to the utility is depicted in Figure 6.1. Because solar PV generates a very low voltage, which is not sufficient to satisfy the inactive and load requirements, respectively, this device consists of a solar PV cell followed by a DC–DC boost converter. The inverter is linked to the output of the converter that was just described. The function of the inverter is to change the DC voltage that is being received into an AC voltage. The transmission rate of the converter is what determines the rate at which energy is converted. Both the current and voltage at the inverter's output phase serve as reference indications for the voltage and current transformation, respectively. The pulse generator is under the control of these two impulses. The input information that is received by those devices is used to determine the control. The main circuit configuration of three phase grid connected system as shown in Figure 6.1.

In order to carry out the simulation, 47 different parallel connected strings are utilized. The voltage throughout the open circuit of the solar PV system is

37.3 V, and the voltage at the place where the power is at its highest is 30 V. The maximum power of the solar PV system is 250.205 W, and each string has a total of 10 series-connected modules. There are 60 individual cells in each section. A photovoltaic system with a voltage of volts has a short circuit current of 8.15 A, shown in Figures 6.2 and 6.3, which depict the characteristics of solar PV cells.

As can be seen in the Figures 6.2 and 6.3, the critical characteristics of a solar PV system are depicted. Figure 6.2 illustrates the relationship between voltage and current, and Figure 6.3 illustrates the relationship between voltage and electricity. The photovoltaic (PV) system is capable of generating a maximum power of 1.176×10^5 W at 360 V when exposed to irradiations of 1000 W/m^2 and a temperature of 24 oC. However, this maximum power decreases to 9.395×10^5 W when exposed to 800 W/m^2 and reaches 5.844 when exposed to 500 W/m^2. The graphs of current and voltage both show evidence of the same phenomenon at the same points in time. When operating at maximal irradiations, the PV system is capable of supplying a load current of 307 V; when operating at second condition, this value is decreased to 305 A; and when operating at third condition, this value reaches 190 A, respectively. The output voltage of the converter as shown in Figure 6.4.

Figure 6.2: SPV-PV characteristics.

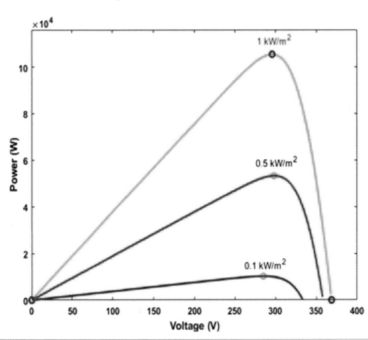

Figure 6.3: SPV-IV characteristics.

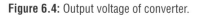

Figure 6.4: Output voltage of converter.

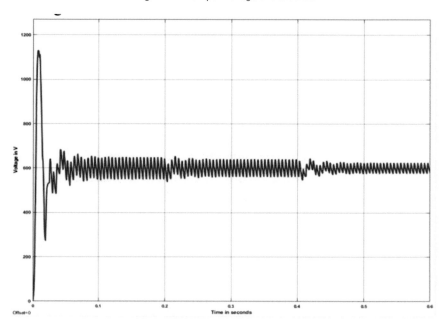

Figure 6.5: Three phase grid connected.

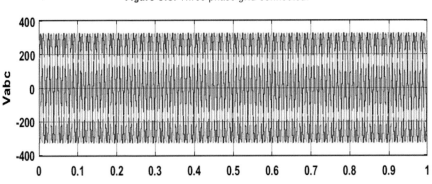

The relationship between voltage and duration is illustrated by Figure 6.5. The boost converter does not produce a DC voltage that is completely unadulterated as shown in Figure 6.5. A significant voltage spike happened very early on in the process. The result is passed into the three-phase inverter, whereas the three-phase grid-connected voltage displays a pure, three-phase sine wave as shown in the Figure 6.5, this is because the suggested system incorporates a voltage and current regulation device.

7

Domestic Application

7.1 Domestic Application

1. Power estimation
2. Energy consumption rate
3. Estimate the electricity bills
4. Choice for solar PV panels
5. Selection of battery and its capacity
6. Charge controller
7. Types of loads
8. Environmental location and its effects on solar PV panels.

The above steps are used to estimate the solar energy for a particular application.

7.2 Load Estimation

Table 7.1: Design of solar PV for domestic applications.

Sl.no	Load name	Power rating (W)	Total number of loads	Run time	Cumulative AC loads (W)	Total energy consumption (kWh)
1	Tube light	35	5	8	175	1400
2	Fan	60	2	6	120	720
3	Television	40	1	4	40	160
4	Fridge	135	1	24	135	3240
5	Air conditioning	1350	2	6	2700	16, 200
6	Motor	500	1	2	1000	2000
Total					4170	23,720

7.3 Measurements of the Converter

As can be seen in the Table 7.1, the anticipated amount of electrical energy is 23,720 kilowatt hours. The load determines how accurate an assessment of the inverter's loss can be. Consider that the inverter has a 20% reduction in efficiency. It is necessary to provide the inverter with additional energy in order to make up for this loss. As a result, the battery has the ability to provide power to 23,464. Because of this, the capacity of the inverter that was chosen is 2.5 kWh.

7.4 Dimensions of the Battery

Because of the extent of the battery's discharge, the energy that was previously contained in the battery is currently unavailable. According to the findings of a number of studies, the depth of depletion of a lead-acid battery is equal to 50% of the battery's total stored electrical energy. While a lithium-ion battery has an approximate 90% depth of depletion when it is fully discharged. Estimates suggest that lithium-ion batteries have a greater depth of depletion than other types of batteries. As a result, a lithium-ion battery is the best option. The average everyday demand for power is approximately 2.5 kW. The following equation can therefore be used to provide a rough approximation of the size of the battery: the capacity of the battery is equal to the amount of energy needed per day divided by the discharge level. In terms of kilowatt-hours, the capacity of the battery is 4.16. However, when choosing a battery, you should take into account both the strength and the number of hours it can store energy. As a result, the largest quantity of battery that can be purchased in today's market is 4.8 kWh.

7.5 Solar Photovoltaic Size

The battery is estimated to have lost 0.15 times its estimated rating due to the estimated loss. 0.72 W of power are being lost from the battery. As a result, the solar PV panel size classification ought to take this loss into consideration. The battery is estimated to have a capability of 4.8 kW, but it has experienced a loss of 0.72 kW. Also take into account the loss of electrical components caused by the inverter. The inverter is what facilitates the transfer of the electricity from the PV to the battery. The converter is anticipated to suffer a loss of 4%. As a result, a panel with a capability of 5 kWp is essential. However, because of the impact on the environment, the solar PV panel has a loss of 25% of its potential output. As a result, the everyday loss is 1.25 kWh. As a result, the total power output of the solar photovoltaic screen is 5.125 kWp. Finally, it is important to determine how many hours of power must be generated by the solar photovoltaic device. Take six hours out of each day as an example. The total quantity of power that must be generated as well as the panel rating are what decide the number of panels that must be utilized. Taking into account a panel with a power output of 300 W, this results in the requirement of 17 solar photovoltaic panels in total.

References

[1] J. Cao, L. Zheng, J. Peng, W. Wang, M. K. Leung, Z. Zheng, M. Hu, Q. Wang, J. Cai, G. Pei, and J. Ji, "Advances in coupled use of renewable energy sources for performance enhancement of vapour compression heat pump: A systematic review of applications to buildings," *Applied Energy*, vol. 332, p. 120571, 2023.

[2] T. Hassan, H. Song, Y. Khan, and D. Kirikkaleli, "Energy efficiency a source of low carbon energy sources? Evidence from 16 high-income OECD economies," *Energy*, vol. 243, p. 123063, 2022.

[3] A. K. Thakur, R. Singh, A. Gehlot, A. K. Kaviti, R. Aseer, S. K. Suraparaju, S. K. Natarajan, and V. S. Sikarwar, "Advancements in solar technologies for sustainable development of agricultural sector in India: a comprehensive review on challenges and opportunities," *Environmental Science and Pollution Research*, vol. 29, no. 29, pp. 43607–43634, 2022.

[4] M. S. I. OumKumari, "Factors influencing Nigerian farmers' adoption of solar water pumps for agricultural application," *Journal of Contemporary Issues in Business and Government*, vol. 28, no. 4, 2022.

[5] M. Tutak and J. Brodny, "Renewable energy consumption in economic sectors in the EU-27. The impact on economics, environment and conventional energy sources. A 20-year perspective," *Journal of Cleaner Production*, vol. 345, p. 131076, 2022.

[6] R. Cergibozan, "Renewable energy sources as a solution for energy security risk: Empirical evidence from OECD countries," *Renewable Energy*, vol. 183, pp. 617–626, 2022.

[7] S. Bouyghrissi, M. Murshed, A. Jindal, A. Berjaoui, H. Mahmood, and M. Khanniba, "The importance of facilitating renewable energy transition for abating CO_2 emissions in Morocco," *Environmental Science and Pollution Research*, vol. 29, no. 14, pp. 20752–20767, 2022.

[8]   T. Xu, W. Gao, F. Qian, and Y. Li, "The implementation limitation of variable renewable energies and its impacts on the public power grid," *Energy*, vol. 239, p. 121992, 2022.

[9]   E. Caceolu, H. K. Yildiz, E. Ouz, N. Huvaj, and J. M. Guerrero, "Offshore wind power plant site selection using analytical hierarchy process for northwest Turkey," *Ocean Engineering*, vol. 111178, p. 111178, 2022.

[10]   S. N. Shorabeh, H. K. Firozjaei, M. K. Firozjaei, M. Jelokhani-Niaraki, M. Homaee, and O. Nematollahi, "The site selection of wind energy power plant using GIS-multi-criteria evaluation from economic perspectives," *Renewable and Sustainable Energy Reviews*, vol. 168, p. 112778, 2022.

[11]   G. Msigwa, J. O. Ighalo, and P.-S. Yap, "Considerations on environmental, economic, and energy impacts of wind energy generation: Projections towards sustainability initiatives," *Science of the Total Environment*, vol. 849, p. 157755, 2022.

[12]   B. Zhao, H. Wang, Z. Huang, and Q. Sun, "Location mapping for constructing biomass power plant using multi-criteria decision-making method," *Sustainable Energy Technologies and Assessments*, vol. 49, p. 101707, 2022.

[13]   M. E. Burulday, M. S. Mert, and N. Javani, "Thermodynamic analysis of a parabolic trough solar power plant integrated with a biomass-based hydrogen production system," *International Journal of Hydrogen Energy*, vol. 47, no. 45, pp. 19481–19501, 2022.

[14]   F. Safari and I. Dincer, "Assessment and multi-objective optimization of a vanadium-chlorine thermochemical cycle integrated with algal biomass gasification for hydrogen and power production," *Energy Conversion and Management*, vol. 253, p. 115132, 2022.

[15]   Y.-P. Xu, Z.-H. Lin, T.-X. Ma, C. She, S.-M. Xing, L.-Y. Qi, S. G. Farkoush, and J. Pan, "Optimization of a biomass-driven rankine cycle integrated with multi-effect desalination, and solid oxide electrolyzer for power, hydrogen, and freshwater production," *Desalination*, vol. 525, p. 115486, 2022.

[16]   H. Hu, W. Xue, P. Jiang, and Y. Li, "Bibliometric analysis for ocean renewable energy: An comprehensive review for hotspots, frontiers, and emerging trends," *Renewable and Sustainable Energy Reviews*, vol. 167, p. 112739, 2022.

[17]   M. Li, H. Luo, S. Zhou, G. M. S. Kumar, X. Guo, T. C. Law, and S. Cao, "State-of-the-art review of the flexibility and feasibility of emerging offshore and coastal ocean energy technologies in East and Southeast Asia," *Renewable and Sustainable Energy Reviews*, vol. 162, p. 112404, 2022.

[18] K. Kumar and R. Saini, "A review on operation and maintenance of hydropower plants," *Sustainable Energy Technologies and Assessments,* vol. 49, p. 101704, 2022.

[19] J. Nasir, A. Javed, M. Ali, K. Ullah, and S. A. A. Kazmi, "Capacity optimization of pumped storage hydropower and its impact on an integrated conventional hydropower plant operation," *Applied Energy,* vol. 323, p. 119561, 2022.

[20] K. Gyanwali, A. Bhattarai, T. R. Bajracharya, R. Komiyama, and Y. Fujii, "Assessing green energy growth in Nepal with a hydropower-hydrogen integrated power grid model," *International Journal of Hydrogen Energy,* vol. 47, no. 34, pp. 15133–15148, 2022.

[21] A. M. Peši, J. Brankov, S. Denda, v. Bjeljac, and J. Mici, "Geothermal energy in Serbia–Current state, utilization and perspectives," *Renewable and Sustainable Energy Reviews,* vol. 162, p. 112442, 2022.

[22] M. Soltani, F. M. Kashkooli, M. A. Fini, D. Gharapetian, J. Nathwani, and M. B. Dusseault, "A review of nanotechnology fluid applications in geothermal energy systems," *Renewable and Sustainable Energy Reviews,* vol. 167, p. 112729, 2022.

[23] R. Cunha and P. Bourne-Webb, "A critical review on the current knowledge of geothermal energy piles to sustainably climatize buildings," *Renewable and Sustainable Energy Reviews,* vol. 158, p. 112072, 2022.

[24] https://www.irena.org/

[25] A. Khare and S. Rangnekar, "A review of particle swarm optimization and its applications in solar photovoltaic system," *Applied Soft Computing,* vol. 13, no. 5, pp. 2997–3006, 2013.

[26] A. Zahedi, "Solar photovoltaic (PV) energy; latest developments in the building integrated and hybrid PV systems," *Renewable energy,* vol. 31, no. 5, pp. 711–718, 2006.

[27] H. Ye, Y. Peng, X. Shang, L. Li, Y. Yao, X. Zhang, T. Zhu, X. Liu, X. Chen, and J. Luo, "Selfâpowered visibleâinfrared polarization photodetection driven by ferroelectric photovoltaic effect in a Dion–Jacobson hybrid perovskite," *Advanced Functional Materials,* vol. 32, no. 24, p. 2200223, 2022.

[28] A.-H. I. Mourad, H. Shareef, N. Ameen, A. H. Alhammadi, M. Iratni, and A. S. Alkaabi, *A State-of-the-art Review: Solar trackers, IEEE, Advances in Science and Engineering Technology International Conferences,* 2022.

[29] A. K. Maurya, A. K. Rai, and H. Ahuja, "Comparative analysis of different MPPT algorithms for Roof-top solar PV system," in *IEE International Conference on Automation, Computing and Renewable Systems 2022.*

[30] B. Aboagye, "Investigation into the impacts of design, installation, operation and maintenance issues on performance and degradation

of installed solar photovoltaic (PV) systems," *Energy for Sustainable Development*, vol. 66, pp. 165–176, 2022.

[31]   V. Shah, G. Kumawat, and S. Payami, "Solar powered electric drive-train with integrated multifunctional dual power on-board charger incorporating N-phase SRM," *IEEE International Conference on Power Electronics, Drives and Energy Systems*, 2023.

[32]   A. Chellakhi, S. E. Beid, and Y. Abouelmahjoub, "An improved adaptable step-size P&O MPPT approach for standalone photovoltaic systems with battery station," *Simulation Modelling Practice and Theory*, vol. 121, p. 102655, 2022.

[33]   A. A. Stonier, G. Peter, and S. Iderus, "Relay Switch Matrix based Optimized reconfigurable Solar PV Battery Charger," *IEEE 19th India Council International Conference*, 2022.

[34]   M. Shafique and X. Luo, "Environmental life cycle assessment of battery electric vehicles from the current and future energy mix perspective," *Journal of Environmental Management*, vol. 303, p. 114050, 2022.

[35]   J. F. Peters, "Best practices for life cycle assessment of batteries," *Nature Sustainability*, vol. 1-3, 2023.

[36]   G. Raina, S. Sinha, G. Saini, S. Sharma, and N. Prashant Malik, "Assessment of photovoltaic power generation using fin augmented passive cooling technique for different climates," *Sustainable Energy Technologies and Assessments*, vol. 52, p. 102095, 2022.

[37]   A. Allouhi, S. Rehman, M. S. Buker, and Z. Said, "Up-to-date literature review on solar PV systems: Technology progress, market status and R&D," *Journal of Cleaner Production*, vol. 132339, 2022.

[38]   D. Zhang, "Improving the performance of PERC silicon solar cells by optimizing the surface inverted pyramid structure on large-area mono-crystalline silicon wafers," *Materials Science in Semiconductor Processing*, vol. 138, p. 106281, 2022.

[39]   N. Iqbal, "Characterization of front contact degradation in monocrystalline and multicrystalline silicon photovoltaic modules following damp heat exposure," *Solar Energy Materials and Solar Cells*, vol. 235, p. 111468, 2022.

[40]   E. P. Busso, "From single crystal to polycrystal plasticity: Overview of main approaches," in *Handbook of Damage Mechanics: Nano to Macro Scale for Materials and Structures*, pp. 1251–1276, 2022.

[41]   A. Wardak, "Effect of doping and annealing on resistivity, mobility-lifetime product, and detector response of (Cd, Mn) Te," *Journal of Alloys and Compounds*, vol. 936, p. 168280, 2023.

[42]   S. Tsurekawa, "Electrical activity of grain boundaries in polycrystalline silicon–influences of grain boundary structure, chemistry and temperature," *International Journal of Materials Research*, vol. 96, no. 2, pp. 197–206, 2023.

[43]   A. Liu, S. P. Phang, and D. Macdonald, "Gettering in silicon photovoltaics: A review," *Solar Energy Materials and Solar Cells*, vol. 234, p. 111447, 2022.

[44]   H. A. Kazem, "Effect of dust and cleaning methods on mono and polycrystalline solar photovoltaic performance: An indoor experimental study," *Solar Energy*, vol. 236, pp. 626–643, 2022.

[45]   A. Sohani, "Thermo-electro-environmental analysis of a photovoltaic solar panel using machine learning and real-time data for smart and sustainable energy generation," *Journal of Cleaner Production*, vol. 353, p. 131611, 2022.

[46]   C. Chen, K. Li, and J. Tang, "Ten years of Sb2Se3 thin film solar cells," *Solar RRL*, vol. 6, no. 7, p. 2200094, 2022.

[47]   N. M. Kumar, "Advancing simulation tools specific to floating solar photovoltaic systems–Comparative analysis of field-measured and simulated energy performance," *Sustainable Energy Technologies and Assessments*, vol. 52, p. 102168, 2022.

[48]   S. Sreejith, "A comprehensive review on thin film amorphous silicon solar cells," *Silicon*, vol. 1-17, 2022.

[49]   S. Saravanan and R. S. Dubey, "Study of ultrathinâfilm amorphous silicon solar cell performance using photonic and plasmonic nanostructure," *International Journal of Energy Research*, vol. 46, no. 3, pp. 2558–2566, 2022.

[50]   M. Bouzidi, "Generalized predictive direct power control with constant switching frequency for multilevel four-leg grid connected converter.IEEE," *Transactions on Power Electronics*, vol. 37, no. 6, pp. 6625–6636, 2022.

[51]   S. Åevik, "Techno-economic evaluation of a grid-connected PV-trigeneration-hydrogen production hybrid system on a university campus," *International Journal of Hydrogen Energy*, vol. 47, no. 57, pp. 23935–23956, 2022.

[52]   T. Debnath, A. K. Jain, and A. R. Bhowmik, "Optimal design of grid tied hybrid microgrid for hospital to enhance reliability and sustainability," in *2nd International Conference on Emerging Frontiers in Electrical and Electronic Technologies*, 2022.

[53]   O. Babatunde, J. Munda, and Y. Hamam, "Hybridized off-grid fuel cell/wind/solar PV/battery for energy generation in a small household: A multi-criteria perspective," *International Journal of Hydrogen Energy*, vol. 47, no. 10, pp. 6437–6452, 2022.

[54]   F. Rebolledo, "Performance evaluation of different solar modules and mounting structures on an on-grid photovoltaic system in south-central Chile," *Energy for Sustainable Development*, vol. 68, pp. 65–75, 2022.

[55]   A. Z. Aghaie, H. Shahabi, and N. Yekta, "Techno-economic feasibility analysis of enhancing photovoltaic lighting systems with intelligent

controllers," *9th Iranian Conference on Renewable Energy & Distributed Generation*, 2022.

[56] R. Khezri, A. Mahmoudi, and H. Aki, "Optimal planning of solar photovoltaic and battery storage systems for grid-connected residential sector: Review, challenges and new perspectives," *Renewable and Sustainable Energy Reviews*, vol. 153, p. 111763, 2022.

[57] S. Ajayan and A. Selvakumar, "Metaheuristic optimization techniques to design solar-fuel cell-battery energy system for locomotives," *International Journal of Hydrogen Energy*, vol. 47, no. 3, pp. 1845–1862, 2022.

[58] B. Zou, "Capacity configuration of distributed photovoltaic and battery system for office buildings considering uncertainties," *Applied Energy*, vol. 319, p. 119243, 2022.

[59] M. Szczepaniak, "Use of the maximum power point tracking method in a portable lithium-ion solar battery charger," *Energies*, vol. 15, no. 1, p. 26, 2022.

[60] A. Najmurrokhman, "Solar panel charge controller using PWM regulation for charging lead acid batteries," in *8th International Conference on Wireless and Telematics 2022*.

[61] L. Bhukya, N. R. Kedika, and S. R. Salkuti, "Enhanced maximum power point techniques for solar photovoltaic system under uniform insolation and partial shading conditions," *A Review. Algorithms*, vol. 15, no. 10, p. 365, 2022.

[62] Z. M. Ali, "Novel hybrid improved bat algorithm and fuzzy system based MPPT for photovoltaic under variable atmospheric conditions," *Sustainable Energy Technologies and Assessments*, vol. 52, p. 102156, 2022.

[63] L. Gong, G. Hou, and C. Huang, "A two-stage MPPT controller for PV system based on the improved artificial bee colony and simultaneous heat transfer search algorithm," *ISA transactions*, vol. 132, pp. 428–443, 2023.

[64] B. Kumar, S. Rao, and M. Indira, "Analysis of grid-connected reduced switch MLI with high-gain interleaved boost converter and hybrid MPPT for solar PV," *International Journal of Energy and Environmental Engineering*, vol. 13, no. 4, pp. 1287–1307, 2022.

[65] M. E. Başoğlu, "Comprehensive review on distributed maximum power point tracking: Submodule level and module level MPPT strategies," *Solar Energy*, vol. 241, pp. 85–108, 2022.

[66] V. L. Mishra, Y. K. Chauhan, and K. Verma, "A critical review on advanced reconfigured models and metaheuristics-based MPPT to address complex shadings of solar array," *Energy Conversion and Management*, vol. 269, p. 116099, 2022.

[67] A. I. M. Ali and H. R. A. Mohamed, "Improved P&O MPPT algorithm with efficient open-circuit voltage estimation for two-stage grid-integrated PV system under realistic solar radiation," *International Journal of Electrical Power & Energy Systems*, vol. 137, p. 107805, 2022.

[68] M. Sarvi and A. Azadian, "A comprehensive review and classified comparison of MPPT algorithms in PV systems," *Energy Systems*, vol. 13, no. 2, pp. 281–320, 2022.

[69] C. Yang and K. Youcef-Toumi, "Principle, implementation, and applications of charge control for piezo-actuated nanopositioners: A comprehensive review." *Mechanical Systems and Signal Processing*, vol. 171, p. 108885, 2022.

[70] S. Lim, "A new PID controller design using differential operator for the integrating process," *Computers & Chemical Engineering*, vol. 170, p. 108105, 2023.

[71] A. Banik, "Design, modelling, and analysis of novel solar PV system using MATLAB," *Materials today: proceedings*, vol. 51, pp. 756–763, 2022.

[72] J. R. Albert, "Design and investigation of solar PV fed single-source voltage-lift multilevel inverter using intelligent controllers," *Journal of Control, Automation and Electrical Systems*, vol. 33, no. 5, pp. 1537–1562, 2022.

[73] C. Bash, P. Akram, M. Murali, and T. T.Mariprasath, "Design of an adaptive fuzzy logic controller for solar PV application with high step-up DC–DC Converter," in *Proceedings of Fourth International Conference on Inventive Material Science Applications*, pp. 349–360, 2021.

[74] C. Basha, T. Mariprasath, M. Murali, and S. Rafikiran, "Simulative design and performance analysis of hybrid optimization technique for PEM fuel cell stack based EV application," *Materialstoday: Proceedings*, vol. 52, pp. 290–295.

[75] C. Basha, M. Murali, T. Shaik Rafikiran, and M. Reddy, "An improved differential evolution optimization controller for enhancing the performance of PEM fuel cell powered electric vehicle system," *Materialstoday: proceedings*, vol. 52, pp. 308–314, 2022.

[76] K. R. T.Mariprasath, "Energy efficiency enhancement of solar PV panel by automatic cleaning technique," *International Journal of Innovative Technology and Exploring Engineering*, vol. 8, pp. 3591–3595, 2019.

[77] J.-D. MarÃn-De-La-Cruz, "Design of a high-reduction ratio, doubly interleaved buck converter," *IEEE International Autumn Meeting on Power, Electronics and Computing*, 2022.

[78] M. I. Uddin, K. S. S. Huq, and S. Salam, "Performance analysis of a newly designed DC to DC buck-boost converter," *Asian Journal of Electrical and Electronic Engineering*, vol. 2, no. 2, pp. 38–42, 2022.

[79]   R. Kumar, "Fuzzy particle swarm optimization control algorithm implementation in photovoltaic integrated shunt active power filter for power quality improvement using hardware-in-the-loop," *Sustainable Energy Technologies and Assessments*, vol. 50, p. 101820, 2022.

[80]   P. Santosh Kumar Reddy and Y. Obulesu, "Design and development of a new transformerless multi-port DC–DC boost converter," *Journal of Electrical Engineering & Technology*, vol. 18, no. 2, pp. 1013–1028, 2023.

[81]   L. S. K. R. Padala and P. O. Yeddula, "A nonâisolated switched inductorâcapacitor cell based multiple high voltage gain DCâDC boost converter," *International Journal of Circuit Theory and Applications*, vol. 50, no. 6, pp. 2150–2174, 2022.

[82]   S. Alkhalaf, Z. M. Ali, and H. Oikawa, "A novel hybrid gravitational and pattern search algorithm based MPPT controller with ANN and perturb and observe for photovoltaic system," *Soft Computing*, vol. 26, no. 15, pp. 7293–7315, 2022.

[83]   A. K. Pandey, V. Singh, and S. Jain, "Study and comparative analysis of perturb and observe (P&O) and fuzzy logic-based PV-MPPT algorithms," *Applications of AI and IOT in Renewable Energy. Academic Press*, pp. 193–209, 2022.

[84]   A. Panda, "Recent advances in the integration of renewable energy sources and storage facilities with hybrid power systems," *Cleaner Engineering and Technology*, vol. 100598, 2023.

[85]   T. Hai, "Second law evaluation and environmental analysis of biomass-fired power plant hybridized with geothermal energy," *Sustainable Energy Technologies and Assessments*, vol. 56, p. 102988, 2023.

Index